TRANS

P

INTEGRATING

THE SEARCH FOR TRUTH

First Place Winner, Non-Fiction Book Category

toExcel
San Jose New York Lincoln Shanghai

ALSO BY E.R. CLOSE

The Book of Atma
A History of Soul Evolution

Infinite Continuity
A Theory Unifying Relativity and Quantum Physics

and

The Search for Certainty
An Account of the Path and the Goal
In the Search for Truth

TRANSCENDENTAL PHYSICS

SCIENCE PROVES THE EXISTENCE OF
GOD
AND INTEGRATES
THE SEARCH FOR TRUTH

Edward R. Close, Ph.D.

Transcendental Physics

Cover designed by Jacquelyn A. Close.

This edition published by toExcel Press,
an imprint of iUniverse.com, Inc.

For information address:
iUniverse.com, Inc.
620 North 48th Street
Suite 201
Lincoln, NE 68504-3467
www.iuniverse.com

ISBN: 0-595-09175-X

ACKNOWLEDGEMENTS

A number of people have had a hand, directly or indirectly in the production of this book. The author gratefully acknowledges those mentioned here, and apologizes for the inevitable omissions. There were many friends, colleagues, and students who engaged in helpful discussions, and encouraged continuation of the research that finally resulted in the completion of the manuscript.

First and foremost, I am very, very thankful for the tireless support of my wife and best friend, Jacqui, who discussed, critiqued, encouraged, and generally aided in the details of this work in every way conceivable, in spite of the time it took away from other equally important activities. My son Joshua also deserves mention for his interest in Transcendental Physics and for his belief in the importance of this work.

The use of concepts and proofs from the work of G. Spencer Brown is gratefully acknowledged. Its importance cannot be overstated. The Calculus of Distinctions, introduced and used in this book, is an adaptation of G. Spencer Brown's Calculus of Indications. While a calculus suitable to describing the logical processes linking consciousness and the physical universe probably would have been developed sooner or later, the far-reaching conclusions reached in this book would not have been possible, or would have remained speculative, until such a calculus was developed.

My long-time friend and college room mate, David Stewart deserves special mention. I bent his ear with some of these ideas as long ago as 1956, and many times since. For their helpful review of the manuscript, I gratefully acknowledge Dr. Amit Goswami, Professor of Physics at the University of Oregon's Institute of Theoretical Sciences, Dr. Ray Knox, Professor of Geoscience at Southeast Missouri State University, and fellow Mensan, Frank G. Pollard, P.E.

TABLE OF CONTENTS

FIGURES

TABLES

FOREWORD

I once heard a minister say, "God is an affirmation of faith, not an inference from fact." That statement is no longer true. This book could be the most important book of the twentieth century, having more long-term impact on the advancement of the human race than any other. It is the logical integration of every valid scientific principle to date and proves that our universe is both physical and non-physical.

Science and religion both claim to address truth--the ultimate realities of our universe. Yet to most people's perception, science and religion are at odds to one extent or another. Religion, which is an intimately personal matter, seems to deal with topics not amenable to the scientific method while science seems to have restricted itself to the impersonal, so-called objective reality--topics outside the fundamental tenets of faith.

Science, to this point, has assumed the universe to function independent of the will and thoughts of its observers while religion, while believing in prayer and a personal God, has asserted that the universe functions, at least in part, in response to the will and thoughts of its observers. When miracles appear to happen, religious followers believe them to be miracles and say "there is no scientific explanation", while scientists do not believe them to be "miracles" saying things like "it was only an illusion, mass hysteria, it never happened, or there is a perfectly logical explanation within the current scientific paradigm, just not yet discovered." While scientists do not accept any reported phenomena as real unless it fits their current paradigm, religionists are willing to accept many types of observations as "beyond science."

According to the author of this book, both are wrong. Miracles can and do happen and do so within the laws already accepted by scientists. It is just that neither scientists nor theologians have come to realize it. This is the book that all scientists in sincere search for the truth and all seekers in search of a reality called "God" have been waiting for--or at least, should have been waiting for.

Dr. Close has accomplished what no one has ever done before. He has taken all the fundamental discoveries and facts of modern science and has proven, conclusively, that none of these laws of science could be true without the all-pervading consciousness of a God--be it called Allah, Brahma, Jehovah, Great Spirit, Yahweh, Heavenly Father, or Divine Mother. In other words, if you accept the laws of modern physics as proven over and over by countless tests and engineering achievements--then you have but one logical conclusion to be drawn, namely that there is a God, a creator that preceded the physical universe, a consciousness that permeates everything and everyone, of whom we are a part, and who knows, is conscious of, and responds to us personally.

With the publication of this work it is no longer logically consistent to be both a scientist and an atheist, nor is it logically correct to believe in God and say that science does not support such a belief. The marriage between science and religion is complete in this book. No longer is there a contradiction. If you believe in one, you necessarily believe in the other, even if you don't realize it. One is not possible without the other. Thanks to this life-long labor by Dr. Close, we can now wholeheartedly practice our faith in God and pursue our science with equal fervor without any possibility of a contradiction or conflict.

He has shown us by exquisite and flawless logic, without passing over or ignoring any aspect of science, that to be a scientist without a belief in God is a contradiction and that it is equally a contradiction to be a spiritual devotee in search of God without accepting the discoveries of science.

One cannot place too much importance on the genius of this work. I, for one am thankful for Ed Close's devotion and commitment to this project that have never flagged over the years. Because of his perseverance, a door has been opened for us all. He has broken forever the barrier between science and religion. What seemed to be antagonists he has made friends--and even better: he has proven them to be one.

Only history will tell what this work will do for the advancement of science and humankind, but its potential is without limit. Science was confined in a self-imposed cage with the majority of the universe on the outside. Ed Close has broken the bars and forged a key link in the chain of advancement of mankind for the ages. This book should be read by everyone who seeks to know the truth. It is worth every effort to study and comprehend--even if you have to read it several times. It could change your entire outlook on life forever, and change it for the better.

To scientists and theologians alike, I say: take this work to heart, for through it you "shall know the truth, and the truth shall make you free." (John 8 : 32)

David Stewart, Ph.D.,
Author, Educator, and Consultant,
Former Director, Central U.S. Earthquake Consortium.
Southeast Missouri State University

PREFACE

Many of the problems of our modern world can be traced to the loss of meaning caused by an increasingly materialistic world view. The exaggerated portrayal in movies and on TV of man's control of the natural world through science and technology has led many to believe that the body is a machine, the mind a computer, and that when the machine wears out, or is irreparably damaged, and the electric impulses of the brain flicker out, nothing remains. Violence, one result of this impoverished view of consciousness, is rampant in the civilized world today. But this need not be the case.

There is ample evidence that most of the really great minds of science, Einstein, Schrödinger, Newton, and De Broglie, just to name a few, were deeply spiritual and religious in the true sense of the word. And now there is incontrovertible empirical evidence that matter and the physical forces associated with it are just the tip of the iceberg of reality. In the pages to come, we will see that this reality, discovered by the deepest probings of science, has all the characteristics of consciousness as we know it, but on a much vaster scale.

As we dig deeper and deeper into the evidence, the nature of this vast form of consciousness will begin to come into sharper focus. Is it the Godhead of Christian theology, the Holy of Holies of the Jews, the Brahman of Hindus, the all-pervading Spirit of the Pantheists, or the Blissful Void of the Buddhists? Is it all of them, or none of them? You will have to make that assessment for yourself. I can only hope that I am able to convey to you the wonder and joy I have found hiding behind the equations, apparatus and minds of the scientists who are striving to understand the nature of our universe.

At any rate, I sincerely hope you will enjoy reading this book as much as I have enjoyed writing it.

INTRODUCTION

The Nonlocal Revolution -- Scientific Proof of the Existence of God

What if a scientific experiment produced indisputable evidence of the existence of God? Could either science or religion accept it?

Scientific proof must be based on unassailable logic; it must produce repeatable empirical results. Such an experiment would have to enable us to determine whether the reality we experience requires a god (i.e., the functioning of a creative consciousness in the formation and perpetuation of that reality), or does not require a god but has evolved, and continues to evolve, independent of any form of conscious activity. The experiment must be repeatable and the determination must be verifiable.

Can science devise such an experiment? Most of us have been convinced that a scientific proof of the existence of God is impossible, and it would certainly be unlikely that a scientist proposing such an experiment would be able to find much serious support among his peers.

On the other hand, if he were proposing an experiment whose sole purpose was to try to determine which theory, relativity or quantum mechanics, more appropriately describes the reality we experience, it is more than likely that he would find a great deal of support. And if that experiment proved empirically that quantum mechanics is the best description of reality, the results of that experiment would warrant scientific jubilation, write-ups in scientific journals, and even mention on the major national and international news networks.

However, if that experiment also provided indisputable, empirical evidence of the functioning of a pervasive form of consciousness, which might be construed to support the possibility of the existence of God, what would happen then? Such an idea is nothing short of revolutionary. Would scientists be eager to break the news of such a discovery to the world?

Like major earthquakes, scientific revolution takes us by surprise, and it may take years for the world to understand exactly what happened. For more than a dozen years now, we've had clear evidence that a mind-boggling scientific revolution is underway. The results of a remarkable experiment, performed in 1982 by the French physicist Alain Aspect and his colleagues, sent a major shock wave reverberating through the entire body of theory and knowledge that make up the current scientific paradigm. The events that led up to this unprecedented revolutionary experiment are traced chronologically as follows:

1900	Planck discovers that energy always occurs in exact multiples of a basic unit (quanta).
1905-1915	Einstein publishes the special theory of relativity, a paper explaining the photoelectric effect, and the general theory of relativity.
1925-1930	Heisenberg develops quantum probability matrices and the uncertainty principle. Schrödinger develops the quantum wave equation. Bohr establishes the Copenhagen interpretation of quantum theory.
1935	The Einstein-Podolsky-Rosen (EPR) paradox and the Einstein-Bohr debate splits physical science into two conflicting camps.
1964	Bell's theorem reveals a way to resolve the Einstein-Bohr debate.
1969	George Spencer Brown publishes the calculus of indications in his book Laws of Form.
1978	John A. Wheeler proposes the delayed-choice, double-slit experiment which is actually performed later by teams in the

US and Germany, independently.

1982 The Aspect experiment validates the Copenhagen interpretation of quantum mechanics. (The results of this experiment have since been repeated and thus verified by a number of experiments.)

1996 Transcendental Physics proves that a pervasive form of consciousness not only exists, but necessarily preceded physical creation.

While Christians call it God, Jews say it is the unspeakable name, *JHVH* (transliterated: Yahweh or Jehovah), and Muslims call it *Allah*. Hindus call it *Sat* and *Parambrahma*--the incomprehensible eternal substance. Buddhists call it *Dharmakaya*: the transcendental Essence or Nothingness out of which all forms spring. Whatever the name or the attributes, every major religion in the world accepts the existence of an Original Conscious Awareness which preceded the existence of physical reality, and pervades all reality. And now, physical science has provided objective proof that they are right!

The current paradigm of scientific materialism starts with the following basic or *a priori assumptions*:

1.) the universe came into existence by chance, requiring neither a creator nor a pre-existing intelligence, and
2.) that the objective world exists independent of the perceiving subject.

The very science that embraces this paradigm has nevertheless produced evidence of a pervasive form of consciousness that exists and operates within physical creation and which preceded the creation of the physical universe. Mathematical proof of the necessity of the existence of such a form of consciousness is presented in this book. In other words, science has proved the existence of God. How could this have happened? Let's look a little more closely at the story sketched briefly in the chronology above.

The 1935 debate between Albert Einstein and Niels Bohr about the true nature of reality spawned a conflict between relativity and quantum physics, the two pillars of modern physical science. In 1964, the Irish physicist John Bell, in a paper that has since become known as Bell's theorem, proposed an experiment that would determine once and for all who was correct, Einstein or Bohr. And, in 1982, the Aspect experiment proved empirically that Einstein was wrong and Bohr was right. Quantum phenomena must register in a way that can be observed and measured before particles and waves can be said to exist as physical objects.

At about the same time that Bell published his theorem (1964) proposing the way to resolve the

Einstein-Bohr debate empirically, an English mathematician named George Spencer Brown published a book called Laws of Form (1969). It contained a wealth of insight into the future direction mathematics must take in order to incorporate the understanding that would be gained from the Aspect experiment, thirteen years later, in 1982. And John A. Wheeler's delayed-choice, double-slit experiments (1978) added yet another piece of the giant puzzle that has taken nearly a century to fit together into a coherent picture.

Yet, finally, it was the Aspect experiment that provided the vital evidence which, when combined with Brown's and Wheeler's work, led to the irrefutable fact that both of the time-honored *a priori* assumptions mentioned above are absolutely wrong. The universe did not come into existence by chance, and the objective world does not exist independent of the perceiving subject. This discovery will be shown to integrate relativity and quantum physics into an internally consistent new paradigm that will revolutionize science. In this book we will tell this story in detail and explain clearly why and how it proves the existence of God, and reveals a great deal concerning the interaction of mind and matter.

So, why haven't you heard about this astonishing discovery?

It is because scientists, like everyone else, want to be certain that their pronouncements are correct. Making a discovery in science can be likened to finding yourself standing precariously on a tall fence. Leap or fall to one side and you are freed of all of your limitations, you experience an epiphany--an ecstasy, and the universe becomes new and vibrant and full of unexpected opportunity. But leap or fall to the other side -the wrong side- and you may drown in a sea of insanity, or be forever lost in the barren wastes of ridicule, berated by your colleagues, friends and foes alike. How, then, when standing on the fault line of a major paradigm shift, does a scientist determine which way to jump?

Some seem to rely on gut feeling. They say they "just know" which way to go. Some cross their fingers, or perhaps say a silent prayer before they take the plunge. And others try to figure the odds. Using whatever means available, they apply the laws of chance, choose, leap, and hope.

The scientific approach, however, is to remain balanced on the fence until there is proof. A scientist weighs and measures. He waits and watches and analyzes the evidence. Which argument is more conducive to objective tests and verifications? Which side amasses the greatest amount of objective proof? Which side is safest?

After sufficient analysis, he elects what appears to him to be the only proper and appropriate choice, and he believes wholeheartedly that his conclusion is the only correct one. When his judgment is questioned, he presents his arguments and, if need be, searches for ways to bolster those arguments. If, at some point in time, his judgment is proved wrong, either logically or empirically, he may take years to examine that proof, but if he is a rational being, he will eventually have to accept it. He has no other choice.

When the entire scientific community has accepted as undeniable an *a priori* assumption and that assumption is proved wrong, then everything that science has built on that assumption comes into question again.

With the Bell's theorem and the Aspect experiment, we have found and climbed atop a new fence. Individuals within the scientific community test, correlate, and corroborate evidence for and against the discovery with all the zeal and dedication of an obsessive-compulsive personality, but the community as a whole waits for the massive body of evidence to tell them which way to jump off the new fence. When irrefutable logic and empirical evidence combine in such a way that the direction to leap is made clear, either the old paradigm is enriched and improved, or it is discarded and a new scientific paradigm is born.

In Transcendental Physics, we will explore the events that have led to the birth of a new paradigm. We will examine the evidence, the logic, and the effects of this monumental discovery on the existing scientific paradigm. We will show, beyond doubt, that an original, non-physical, Primary Consciousness had to exist prior to the first quantum of physical matter. Then, we will turn our searchlight upon the new frontiers that await scientific exploration, and upon the powerful, positive, stimulating implications this discovery offers to humanity.

If you share the doubts of the majority of the modern world, and think that a scientific proof of the existence of God is impossible, all I ask is that you sit on the fence with the prudent scientist and withhold final judgment until you have examined all the evidence thoroughly. That evidence is presented in this book.

CHAPTER 1. THE EINSTEIN FALLACY

No elementary phenomenon is a phenomenon until it is a registered phenomenon.

- John Archibald Wheeler summarizing Bohr's response to Einstein.

If we compare the course of science to a river, it appears as a surging stream of thought, flowing from the springs of the great minds of the past, toward the oceanic goals of universal knowledge and understanding. The turbulence of its escape from the valley of religious dogma nearly forgotten, it flows fairly smoothly out of the last century, only to be churned into foaming rapids at the conjunction of the two powerful tributaries, relativity and quantum mechanics.

Downstream, we hear a deafening roar, and billowing mists warn of a great precipice, an impending plunge over a giant waterfall. The impetus of the clash of the two great tributaries drives us inexorably toward the frightful, yawning brink. Some of us, seeing only chaos and confusion, fear for the destruction of our carefully constructed theories. Others among us seeing great beauty in the wild scene, are too stunned to be concerned about the coming turmoil, too much in awe of the power of the new ideas. A closer look at the barque in which we scientists are travelling tells us that the creaking hull of the current paradigm will probably not survive the pounding of the rapids leading to the falls. A great crack is opening even now and the swirling waters of new concepts are engulfing us. But downstream, beyond the falls, we can see a glistening rainbow arching through the mists above an even more powerful river that flows on, toward the tranquil Sea of Enlightenment.

Advances in science come about in two very different ways. The first is sure and comfortable: progress by the careful process of building, step by step, upon what is known. The second requires a revolution.

Most scientists, like most people, cling to what is sure and comfortable. They are wary of revolutionary ideas and often with good reason. Why risk professional embarrassment or ridicule when they don't fully understand a new idea?--especially if they fear the possible implications or repercussions of that idea. Is it possible though, for a true revolution to go unnoticed or perhaps be ignored?

We find ourselves at a critical point in the history of science and human thought, a point where it is necessary for us to break away from old, out-moded ideas.

So, how did we get to this point? With the success of Kepler and Newton's celestial mechanics, it seemed certain that science was on the right track. Thinkers like Simon La Place and Lord Bertrand Russell saw the universe as a machine; a vast system of atoms, molecules, planets and stars, but a machine none the less, and with the mechanics understood, its past could be known and its future predicted. All scientists had to do was work out the details. Unfortunately, while pursuing that goal, certain anomalies came to light, such as the strange orbits of electrons and of the planet Mercury. Einstein, delving deeper into Maxwell's wave equation and the Lorentz contraction equations, saw the relation between mass-energy and space-time and developed the theory of relativity, which explained some of the anomalous observations that had puzzled classical physicists for many years.

For a time, it appeared that relativity was little more than a minor correction to Newton's mechanics, negligible except at the extremes, and physicists were content with the progress, even though many were still uneasy with the theoretical baggage that the theory of relativity brought with it. Gone forever was the universal space-time reference frame. The concept of aether was replaced by a much more complex, relative space-time continuum that was motion sensitive and could be bent and warped by extreme mass. Relativity said that events that are simultaneous for one observer, might not be so for another, and no event can be said to occur at a unique point in time or space. The location and sequence of a series of events can only be described relative to an observer, which means that the history of any series of events may be different for different observers. And, even stranger still, according to the theory of relativity, distances shrink, time slows down, and mass increases as relative motion increases.

Yet, even as relativity was gaining acceptance, quantum mechanics began its ascendance. Starting with Planck's discovery that mass and energy are always observed in discreet units, or quanta, Bohr, Schrödinger, and Heisenberg developed this strange new science with religious zeal. Using quantum theory, Bohr was able to explain how stable atomic structure could exist, contrary to the concepts of classical physics. Quantum mechanics resolved many problems in particle physics and subsequently revolutionized micro-electronics. It was just as successful, in its own way, as was relativity. Yet it also brought along its own theoretical baggage: Waves and particles had to be treated as complementary aspects of matter and energy. Sub-atomic particles could no longer be thought of as behaving like little billiard balls. And energy, even in waves or fields, was observed to react in finite increments, or quanta. Perhaps the biggest fly in the ointment for deterministic physicists was quantum uncertainty. Heisenberg's uncertainty principle said that either the location or the angular momentum of a particle could be determined with certainty, but not both, because the measurement of one introduced an element of uncertainty in the determination of the other.

And of all scientists, it was Albert Einstein who found quantum theory the most objectionable. A deeply religious man in the universal sense, he once said he could not believe that God would play dice with the universe. Einstein was a determinist. He spoke of the laws of nature as God's thoughts. He saw the universe as a complex but none the less finite physical system, designed and created by God, in such a way that the mechanics could be discovered by science, the past understood and the future predicted. The basic assumption of scientific materialism lay at the heart of Einstein's science, as it did for virtually all of his colleagues.

In James Clerke Maxwell: A Commemorative Volume (Cambridge University Press, 1931) Einstein expressed this bias clearly when he said: "The belief in an external world independent of the perceiving subject is the basis of all natural science."

With this belief in an external world that could be objectively examined and quantified firmly entrenched in their minds, many scientists found it difficult to accept the Copenhagen interpretation of quantum mechanics. And there is little surprise in this fact when one understands that the Copenhagen interpretation tells us that no quantum particles or waves exist apart from the apparatus of observation, and no quanta exist until they register in such a way as to be measurable and observable.

Einstein is said to have asked a colleague: "Do you really believe that the moon only exists when you are looking at it?" Like Dr. Johnson, kicking the stone to refute Bishop Berkeley's subjectivism, Einstein was appealing to common sense in an effort to refute what he saw as a new form of the same sophistic argument. On another occasion, he wrote to a friend: "Even the great initial success of the quantum theory does not make me believe."

The Copenhagen interpretation is not easy to accept and understand. It takes some getting used to. Yet, in 1982, Alain Aspect and his colleagues proved beyond doubt that the Copenhagen interpretation is correct.

How was it possible that these two highly successful scientific disciplines, relativity and quantum theory, started with the same basic assumption that reality exists independent of the conscious observer and pursuing the same general goals, still came into conflict?

In a paper entitled "Can the Quantum Mechanical Description of Physical Reality be Complete?" published in collaboration with Boris Podolsky and Nathan Rosen, Einstein sought to prove the uncertainty principle false by counter example. The line of reasoning presented in this paper became known as the EPR paradox.

Heisenberg's uncertainty principle states that for a given quantum particle, if one of the two basic physical parameters, position or angular momentum, is known, the other can only be determined approximately, within a certain range of probability.

EPR said that we should consider a pair of identical quantum particles travelling in opposite directions after being created by some atomic or subatomic reaction. Due to their common origin and the conservation of mass and energy, they possess certain complementary attributes. If the intrinsic angular momentum (commonly called spin) of one particle is measured, the spin of the other is known, because they are complementary. If such particles were intercepted at opposite ends of a laboratory, EPR argued, by measuring the exact location of one and the spin of the other, both bits of information could be known about the particles, thereby creating a paradoxical contradiction of the uncertainty principle.

Quantum physicists were thrown into a quandary, for a time. Their whole theory seemed about to come crashing down around their ears. After some hard thinking, they countered that the EPR argument was in error because it assumed that quantum particles exist before a measurement is made.

Bohr declared that the error lay in thinking of the particles as separate from the apparatus of observation. Until an observation or measurement is made, the particles exist only as a set of possibilities. In this way, quantum uncertainty is preserved since there is no way of knowing how performing a measurement on one particle will have disturbed the system as a whole. Even though the two bits of information may be obtained by the EPR procedure, Bohr concluded that you cannot say that the two pieces of information ever applied simultaneously to the same particle. This position became known as the Copenhagen interpretation of quantum mechanics, taking the name of the city where Niels Bohr lived and worked.

Einstein's response was that if this fog of probability prior to observation is accepted, it raises another, even more serious problem: Suppose that the particles are intercepted in laboratories set up at opposite ends of the galaxy. If the first particle, with its physical parameters of velocity and spin, cannot exist until a measurement is made, and the particles have complementary spin (as they must because of conservation of momentum) then the first particle will have communicated physical information to its twin across the galaxy instantaneously, transmitting information in excess of the speed of light.

"No reasonable definition of reality can be expected to permit this," he concluded. Einstein had explored the implications of super-luminal velocities while developing the theory of relativity, and he knew that such velocities would wreak havoc with our understanding of time and space.

Neither side would concede to the others' points of view and the debate continued until thirty years after Einstein's death. The resolution came in stages. First, in 1964, John Bell, an Irish-born physicist working in Switzerland, produced a mathematical demonstration (see Appendix A) showing that if accurate measurements were made on a large number of pairs of twin particles, like electrons or photons, there would be a significant difference in the results, depending upon whether reality is local, behaving as Einstein believed it should, or nonlocal, as the Copenhagen interpretation of quantum theory predicts. This mathematical demonstration became known as Bell's theorem. The importance of this work can hardly be overestimated because, as a mathematical theorem, it has been proved. This distinguishes it from a theory, such as scientific materialism, which is simply a working hypothesis.

Bell's theorem began with a carefully designed version of the EPR argument. The twin particles shoot off in opposite directions. See Figure 1. If the particles are electrons, only two orientations of

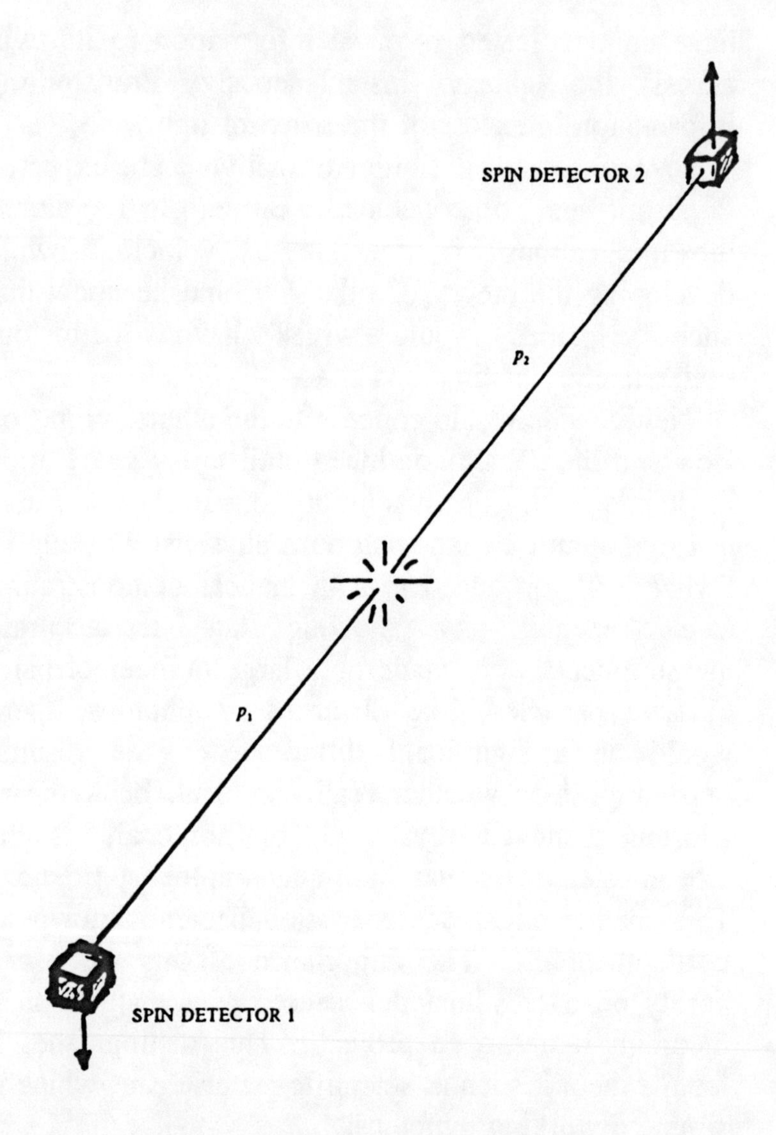

Figure 1. Bell's Theorem Twin Particle Experiment

intrinsic angular momentum, or spin, are possible. By convention, they are called "up" and "down" spin. There is no way of predicting for any given pair, which twin will have one spin and which twin will have the other. At some distance from their point of origin, they are intercepted by devices called spin detectors, which force the particles to reveal their spin by sending particles with "up" spin along one path, and particles with "down" spin along another. By rotating the spin detectors with respect to each other, the perfect correlation between the particles of each pair could be disturbed.

Bell showed that when the spin detectors were 45° out of parallel, the correlation between quantum pairs in a local reality, travelling as objective particles, must lie between -2 and +2, while the same correlation for quanta in a Copenhagen interpretation reality must be equal to 2.83. (For a more complete explanation of how these numbers are derived, see Appendix A.) Thus, quantum physics predicted a correlation that was not only significantly different from the correlation predicted by relativity, but one with a specific value that would be impossible to obtain in the type of local reality posited by relativity and classical physics. Because the experiment suggested by Bell's theorem was very difficult to set up and perform accurately, it was almost 20 years before indisputable results were achieved.

A number of experimental scientists devised instrumentation to produce twin particles and accurately measure their physical parameters, and even the early experiments produced results that appeared to favor the Copenhagen interpretation. The more precise the experiment became, the closer the correlation came to the Copenhagen interpretation prediction. Finally, in 1982, an experiment carried out by Alain Aspect and his colleagues in France, using photons as the particles, yielded unequivocal results. (See Appendix B.)

There could no longer be any doubt: Elementary particles do not behave as Einstein thought they should. Although they appear, after the fact, to have traveled along definite paths from their source to the receptor, neither they nor their paths actually exist until they have impacted on a photographic plate, a collector, a particle counter, i.e., until they have produced an effect that can be observed and measured.

Conclusion: **Einstein was wrong. Bohr's Copenhagen interpretation is correct.**

Verification of the Copenhagen interpretation of quantum mechanics as a valid description of reality is an event of immense scientific importance. But many scientists are still very uneasy with some of the implications. The Aspect experiment has proved that the particles believed to be the basic building blocks of physical reality do not exist until they have impacted on a receptor so that an observation or measurement can be made. Another way to say this would be: "No elementary phenomenon is a phenomenon until it is a registered phenomenon." This is something totally new that has been added to our understanding of the nature of reality, and it is profoundly at odds with the basic assumptions of the current scientific paradigm.

Niels Bohr was certainly aware that his theory was a departure from what had gone before. He was quoted by Heisenberg (Physics and Beyond, Harper and Row, New York, 1971) as follows: "...those who are not shocked when they first come across quantum theory cannot possibly have understood it."

Based on the Bell's theorem-Aspect experiment outcome, some physicists have predicted a paradigm shift rivaling or even surpassing the shift from classical physics to relativity and quantum physics. Physicist Henry Stapp called Bell's theorem "the most profound discovery in science." And Michael Talbot, in Beyond the Quantum (Macmillan, 1987) had this to say:

> Perhaps the most startling feature of the Aspect experiment is that such a historic event took place with so little notice. Immediately following the experiment, many physicists had remarkably little to say about Aspect's dramatic disclosure. Between 1982 and 1985, a few articles appeared in scientific journals praising the work of the French team, but assessed it ultimately with only brief and tantalizingly vague sentences like "lead(s) to realities beyond our common experience" and "indicate(s) that we must be prepared to consider radically new views of reality"...

The implications of the Copenhagen interpretation are so profound, they bear repeating: Physical, subatomic particles do not exist before producing observable effects, nor do they exist as objects independent of the apparatus of observation.

The fact that observation can have effects on physical phenomena has been known to physicists for a long time. It is easy to demonstrate that light can be forced to manifest itself as either wave or particle, depending upon a choice made by an observer, by performing a simple variation on the well-known two-slit experiment. (We will examine this experiment in detail in Chapter 2.) But the real question becomes: <u>Is observation necessary for the objective formation of physical reality</u>?

As I sat thinking about the Copenhagen interpretation of quantum reality, wondering whether mere observation could bring physical energy and matter out of some strange probabilistic state into concrete manifestation, (and if so, how?) a hummingbird visited a pot of flowers hanging outside my window. He paused, hovering before the glass, as if looking in at me. This iridescent little bundle of energy must have had a history prior to my seeing him, I thought. There must have been a tiny nest and tiny parents. The remnants of a tiny egg must be degrading somewhere in the acres of tall oak and hickory trees that surround my home. There certainly was an awareness behind the sparkle of those tiny jewel-like eyes, a certain level of consciousness. Is it possible that observations made by a hummingbird can cause the collapse of probabilistic wave functions into the electrons, protons, and atoms that form the physical, feathered body darting about outside my window? Perhaps. But what about inanimate objects? What of Einstein's moon? Surely the very moon that I might see tonight via a stream of newly-born photons, has a history marked on its crater-scarred face consistent with the moon that Einstein pointed to more than half a century ago. Yet the views of Bohr and Heisenberg, comprising the Copenhagen interpretation, were proved correct by Bell's theorem and the empirical evidence of the Aspect experiment in 1982. Photons, electrons, and other high-speed sub-atomic quanta can not be said to have localized physical form or definite trajectories through space until a measurement or observation is completed.

Classical physics was based on the assumption that objective reality consists solely of matter and energy interacting over time in empty space. While matter and energy were understood to be subtly interwoven, they were believed to be categorically different. This belief was abruptly overthrown by Einstein's theory of relativity. Now we know that matter and energy are not categorically different. There are natural processes by which they are transformed one to the other, matter to energy and energy to matter. Einstein also demonstrated that, contrary to the notions of classical physics, time and space are not changeless backdrops whose dimensions are independent of matter, energy, and human observation.

With the advent of quantum mechanics, it became even clearer that matter, energy, time, and space are all inextricably woven together in the fabric of reality. And now we have strong evidence that observation is required for reality to manifest as matter and energy. At the very least, the results of the Aspect experiment imply that there is no way to prove that quanta of matter and energy exist prior to the completion of an observation. But what is the mechanism by which all observation is completed?

In spite of the fact that science has embraced the basic assumption of the absolute separation of subject and object, mind and matter, the conflict between relativity and quantum mechanics has produced results that challenge the belief in such an independent reality, and we now have hard evidence that reality is not what we have assumed it to be.

Our natural curiosity will not allow us to turn away from the door through which we have glimpsed such fascinating new aspects of the universe we experience, because a deep-seated desire to <u>know</u> underlies all forms of inquiry. We want to understand everything we can about ourselves and the circumstances that make our existence possible. Indeed, science represents mankind's most sustained effort to date to satisfy this innate curiosity.

The results of the Aspect experiment suggest that the vast structure of the universe, with everything it contains, including sentient life forms, consists of an effectively infinite tableau that is far too rich and varied to be circumscribed in simple, materialistic terms. Yet the scientific establishment still wavers in uncertainty, somewhere between denial and acceptance, while it waits for further evidence.

As physicists have begun to try to deal with the results of the Aspect experiment, more books, articles, scientific papers, and essays are being written in attempts to explain away the paradoxes and conflicts inherent in the current relativity-quantum paradigm. Individuals and teams of scientists are hard at work, formulating new theories and combinations of old theories in efforts to reunite physical science. The mathematical and conceptual sophistication of the resulting theories, from David Bohm's hidden-variable, implicate-explicate orders of reality, to the Yang-Mills gauge theory, to the latest superstring theories, all are truly staggering. Multi-dimensional universes, parallel universes, and hyperspace riddled with black holes and wormholes abound, but paradox and conflict still persist.

Physicists agree that the Copenhagen interpretation is correct, but there is very little agreement about exactly what it means. The majority believe that the physical arrangement of equipment (e.g., a camera or photographic plate) can cause a particle (e.g., a photon) to manifest without the involvement of a conscious observer. But there are a few who insist that the process of <u>conscious</u> observation is ultimately necessary for the objective manifestation of particles and/or waves. Perhaps foremost among them is the great Hungarian mathematician John von Neumann. His definitive work: <u>The Mathematical Foundations of</u>

Quantum Mechanics, (see pages 417-445, Princeton University Press, Princeton, NJ. 1955, translated from the German edition by Robert T. Beyer) was written around 1928 to 1930, and is considered by many physicists to be the "bible of quantum mechanics."

In the pages cited above, von Neumann shows that there is no reason not to treat all physical systems as quantum systems. Furthermore, he demonstrates that all the statistical properties of the physical systems involved in an observation may be derived from their wave functions and fully and consistently described using Hermitean operators in Hilbert space (see glossary). Using this mathematically rigorous approach, he demonstrates that there is no functional difference between separate and composite systems, and no physical mechanism for wave collapse. Thus the consciousness of the observer is the only factor left to bring about the collapse. The inescapable conclusion that reality is created by the action of consciousness was picked up and developed by physicists Eugene Wigner, Fritz London, Edmund Bauer, and others. But for every physicist who holds this view, there are dozens who vehemently oppose it.

Clearly, we may have a propensity to believe one way or the other, but we cannot afford to answer this question compulsively on the basis of belief or assumption, however tempting that might be. We must instead insist on further investigation. Is conscious observation the cause of physical reality, or would the same physical reality exist with or without the presence and involvement of a conscious entity or entities? Do the results of the Aspect experiment indicate which is the true picture of reality? Is there other information favoring one conclusion or the other? We will examine the evidence.

There can be only one reality, and everything from quark to galaxy, from amoeba to scientist, exists within it. Bell's theorem and the Aspect experiment have proved that our one and only reality is nonlocal, and the question has been raised concerning the involvement of consciousness in the formation of physical objects. If the functioning of consciousness proves to be the cause of the particle and wave nature of matter, science will never be the same. The whole theory of the origin of man and the universe will have to be rewritten. But even if consciousness proves to be an effect rather than the cause of physical reality, we still have to develop a scientific paradigm that can accommodate photons, electrons, and other elementary particles that do not exist until they register in a way that they can be measured and observed.

The new paradigm will also have to explain what time and space are in a nonlocal reality and how a choice made by a conscious observer can determine the form of the physical phenomena observed. Bell's theorem and the results of the Aspect experiment force us to take the Copenhagen interpretation seriously. And the question of the nature of the involvement of consciousness in physical reality needs to be answered. It is, therefore, clear that the world must brace itself for a shift to yet another strange new paradigm. Our understanding of the nature of reality is about to be radically changed forever.

In this book we will review what we know about the interaction of consciousness and matter and offer some ideas that may be used to fashion a new vehicle for moving forward into the exciting new territories revealed by Bell's theorem and the Aspect experiment. We will broaden and expand the scope of physical science to include more than the quanta that underlie physical reality. If successful, this approach may justifiably be described as transcendental physics.

CHAPTER 2. THE FINAL RECEPTOR

The city of Jeddah, in the Kingdom of Saudi Arabia, is one of the oldest cities in the world. Residents of Jeddah will tell you that the first woman, Eve, is buried there. In fact, you can visit Eve's Tomb inside the walls of the Old City Section of Jeddah. Reverence for Allah lies just beneath the surface in this bustling city which is the primary gateway to the holy cities of Mecca and Medina.

Amartech, Ltd., the company I worked for, occupied a three-story building that had been built as a private residence in the As-Salamah District. One day I noticed that a tile near one corner of a wall in the bath room on the third floor was turned 90 degrees out of orientation with all the others. The beautiful blue geometric pattern on the white ceramic tile marched across the wall in precise repetition, except for the one tile. I searched the adjacent wall and found that it too had one tile turned the wrong way. I soon discovered that every wall in the building, and even the floors, each had one, and only one, tile out of alignment. I began to notice other imperfections, that seemed deliberate, in other aspects of the structure. The stairs, for example, had one step different from all the others. I mentioned these observations to Salah, our Egyptian accountant.

"The workmen were good Muslims", he declared. "They know that only Allah is perfect."

Reflecting on this, I realized that absolute perfection is hard to find in the physical world. We think of billiard balls and precision-ground ball bearings as perfect spheres, but if we look at them closely enough, we find pits and ridges somewhat like the craters on the moon. We conceive of perfect shapes and forms in consciousness, picturing perfect smoothness, roundness, and symmetry in our mind's eye, but we can only approximate such shapes in the physical world, no matter how hard we try. Perhaps the whole universe is only an approximation of a thought in the mind of God. Perfect symmetry is only roughly approximated in material objects, more closely represented in energy, electromagnetic waves, for example, even more closely approached in the visions of human consciousness, and perhaps exactly achieved in the mind of God.

Biologists, physiologists, and some physicists are looking hard at the structure and functioning of the brain for mechanisms that might explain the phenomenon of consciousness in terms of the physical forms of matter and energy, but they may be looking in the wrong place. Why have scientists been unable to come to grips with the implications of Bell's theorem and the empirical results of the Aspect experiment? The materialistic bias of the current paradigm, promotes the same fallacy that led Einstein to challenge quantum mechanics: the assumption of the absolute separation of consciousness and matter. The effect of this assumption is the dismissal of consciousness as a real force or substance. Instead of looking for what happens in matter, the new evidence suggests that we should be investigating what happens in consciousness when an observation is made.

A scientific treatment of consciousness is sorely needed. But such a treatment cannot be properly developed as long as we insist that consciousness is not something as real as matter and energy. The idea that consciousness might be able to affect matter directly is considered to be heresy by believers in the current materialistic paradigm. The reversal of this attitude will require nothing less than a scientific revolution.

It certainly is not the case that scientists want to avoid the truth when it doesn't conform to their current theories, but we will not easily give up the theories that we know and have already understood. Educated and trained in an established paradigm, careers built upon its universally accepted framework, scientists have not only an ethical adjunct to carefully scrutinize new ideas and a professional liability that proscribes rash judgements, but also have deeply held beliefs and deeply personal fears that make it difficult to see beyond that established paradigm. We find it rational and logical to denounce anything outside of or beyond the paradigm as nonsense, or even paranormal.

This state of affairs is not, however, unique. It has happened many times before. Galileo is quoted by Alex Comfort in Reality and Empathy: Physics, Mind and Science in the 21st Century (page 24), State University of New York Press, New York, 1984, as criticizing his contemporary, Johannes Kepler's theory regarding tides:

> Among authorities who have theorized about the ... phenomena, I am most shocked by Kepler. He was a man of exceptional genius, ... he had a grasp of terrestrial movement, but he went on to ... get interested in a supposed action of the moon on water, and other "paranormal" phenomena -- a lot of childish nonsense.

More recently, in the summer of 1905 a paper entitled "On the Electrodynamics of Moving Bodies" was published in the German scientific journal Annalen der Physik. In this short paper, an almost unknown patent office clerk outlined an amazing new theory, but it wasn't taken very seriously by the majority of the scientific community at that time. Some of them called it utter nonsense. However, the strength of Einstein's vision was not to be denied, and as a result, our understanding of matter, energy, time, and space was revolutionized. A few years later, Einstein himself, in turn, refused to accept quantum mechanics as a valid description of reality.

Now, another revolution is in the making. An integrated transcendental science of reality is about to blossom forth, and the seeds of this new science lie in the Copenhagen interpretation, Bell's theorem and the Aspect experiment.

Do Bell's theorem and the Aspect experiment prove the existence of an innate form of consciousness underlying all physical reality? Many in the scientific establishment seem very reluctant to explore this possibility. Their reluctance is understandable, since accepting even the possibility of the direct involvement of consciousness in quantum phenomena implies that materialism might not be a valid basis for scientific investigation. The materialistic bias is so ingrained in the scientific community that this is a difficult idea for scientists trained under the current system to understand and accept.

As a result of this bias, proof that the substance of physical reality does not exist as particles or waves until precipitated from the spectrum of probable states by being caused to register as measured or observed phenomena (the Copenhagen interpretation) has not liberated many scientists from belief in materialism. Most physicists find it reasonable to assume that, even if the basic building blocks of physical reality behave in this strange way, the physical universe would exist pretty much as it is, with or without the presence of a conscious observer. They reason that photons, electrons, and other elementary particles should be scattered and refracted by existing physical features, waves of energy meeting in empty space should create interference patterns, and photons should make pin points of light on sensitive surfaces with or without the presence of a conscious entity to observe them.

This point of view is expressed clearly by physicist John Wheeler, on page 126 of At Home in the Universe, American Institute of Physics, 1994:

> "Consciousness" has nothing whatsoever to do with the quantum process. We are dealing with an event that makes itself known by an irreversible act of amplification, by an indelible record, an act of registration. Does that record subsequently enter into the consciousness of some person, some animal, or some computer? Is that the first step in translating the measurement into "meaning"--meaning regarded as "the joint product of all the evidence that is available to those who communicate?" Then that is a separate part of the story, important but not to be confused with "quantum phenomenon."

Wheeler's strong statement, which implies that the universe could exist as we perceive it without the functioning of consciousness, is only an assumption, not a scientific conclusion. In fact, since there is no

way to observe a universe without consciousness, the statement does not even qualify as a scientific hypothesis, --because it cannot be tested.

Certain physical appearances have led scientists to assume that the structures existing in the universe have evolved without the benefit of consciousness, but we have been deceived by appearances before. The sun appears to move around the earth. Physical objects such as metal and stone appear to be solid. How can anyone categorically deny the possibility that consciousness has anything to do with quantum phenomena when we don't even have a good working definition of consciousness? Science within the current paradigm has very little to say about what consciousness is, or why it exists, and only slightly more to say about how it functions.

The appearances that have led scientists to assume that consciousness is an emerging feature of the evolving physical universe arise from the assumption that consciousness is something only associated with higher forms of organic life. We should be wary of this assumption. Consciousness, like matter and energy, may come in many different forms. Our egocentric view that consciousness is a phenomenon associated primarily with the human brain is undoubtedly too narrow. If it turns out that consciousness does have something to do with quantum processes, how far into the structure of the universe does this involvement reach?

The Copenhagen interpretation, coupled with the results of the Aspect experiment may force us to rethink the nature of consciousness and its role in the apparent structure of the universe. In the Aspect experiment, calcium atoms are excited by lasers, causing them to emit pairs of correlated photons. Because of the conservation of mass and energy, they have equal and opposite velocities and identical polarization (analogous to spin). We might ask whether the experiment will yield the same results if elementary particles other than photons are used. The logic of the EPR thought experiment is the same for any pair of identical, correlated particles, and Bohr's statement that elementary quanta do not have paths or exist until they produce effects that can be observed and measured implies that all elementary particles behave this way. A look at the famous two-slit experiment, originally devised in 1801 by an English physician named Thomas Young to demonstrate the wave nature of light, will help us to gain a clear understanding of the Copenhagen interpretation by providing an exercise in thinking about the behavior of elementary particles.

Since the advent of quantum mechanics, the classic double-slit experiment has been modified to show that an observer can cause light to exhibit either wave or particle nature, depending on how he chooses to set up the experiment. Numerous experiments have demonstrated that this quantum behavior is true for all types of elementary particles, and since observations and measurements involve conscious choice, it provides a clue to the way consciousness and matter interact.

The basic two-slit experiment is very simple, consisting of a light source, an opaque plate with two slits in it, and a screen or photographic plate (Figure 2). When we open both slits, bands of light and shadow are created on the screen, displaying the alternating re-enforcement and cancelling effects of the interfering light waves. If we then cover one of the slits, the screen displays a scattered light pattern directly behind the open slit, consistent with the expectation for impacts by individual photons. Thus by opening or closing one of the slits, we can choose to see light behaving as either waves or particles.

The two-slit experiment has been refined in recent times to the point that a single quantum of light - <u>one photon</u> - can be released at a time. A very sensitive photographic plate is used as the screen. Sending the photons through one at a time, with both slits open, experimenters were surprised to find that, no matter how far apart in time the individual photons were released, a typical banded wave-interference pattern developed when sufficient photon impacts had registered on the screen.

How can this be explained? It would seem that either each particle must somehow know the path of

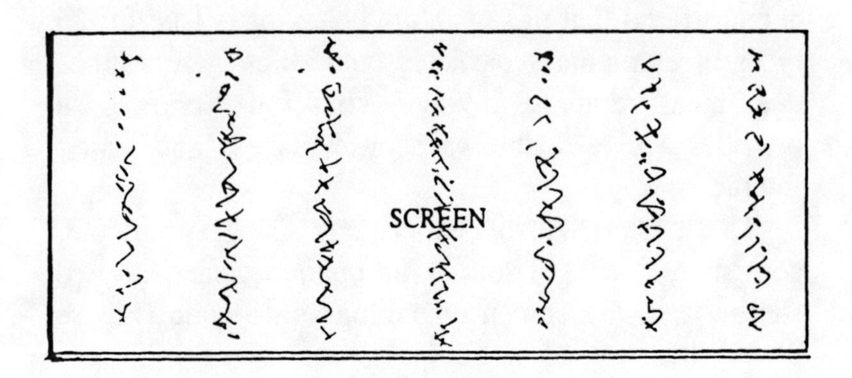

Figure 2. Young's Two-Slit Experiment

every previous particle and use that information to decide where to land in order to contribute to the interference pattern, or each particle manages, somehow, to pass through both slits and interfere with itself. The first possibility seems extremely far-fetched, attributing some sort of consciousness or group memory to elementary particles, and the second possibility appears to be impossible since, by definition, a quantum is indivisible.

This strange behavior is not limited to photons. The double-slit experiment and variations have been carried out many times using electrons and other elementary particles. The results demonstrate that all different kinds of elementary particles behave in exactly the same way. But we are forgetting what Niels Bohr said. Remember the basic tenet of the Copenhagen interpretation: Quanta do not exist until they register in an observation or measurement, and it is incorrect to think of the quanta as having paths existing apart from the whole apparatus.

If photons, electrons, and all other elementary particles do not exist in flight, then perhaps they don't interact with the slits as particles or individual wave packets at all. What will happen if we wait until after a particle has had time to pass the slits and make our decision to observe particle or wave phenomena just before the photon should be arriving at the screen? By 1985, physicists had refined the two-slit experiment specifically to answer this question. The apparatus was arranged as depicted in Figure 3.

The source at A emits one photon at a time and the lens at B is coated with an opaque material except on two vertical lines, creating two slits. The photographic plate at C is composed of slats, like a venetian blind. The slats can be opened to allow the photon to pass on through and be collected by one or the other of the photon counters at D, depending upon which slit it has passed through, or it can be closed to function as a screen to record the pattern of photon impact on the photographic surface.

Consider one specific, single photon, released by the source at A. If the photon behaves as a finite particle in a local reality, it will travel from A to D, passing through one slit or the other. We will know which slit it passes through by noting which photon counter receives it. When the venetian-blind photographic plate is open, this appears to be what happens. When the venetian blind is closed, the photon exhibits wave behavior by striking the photographic plate and contributing to a banded interference pattern.

Now, if the distance between B and C is great enough, we can calculate the time that it will take the photon, travelling at the speed of light, to reach the slits, and open or close the venetian blind after the photon has passed through the slits, yet before it arrives at the location of the venetian blind. This

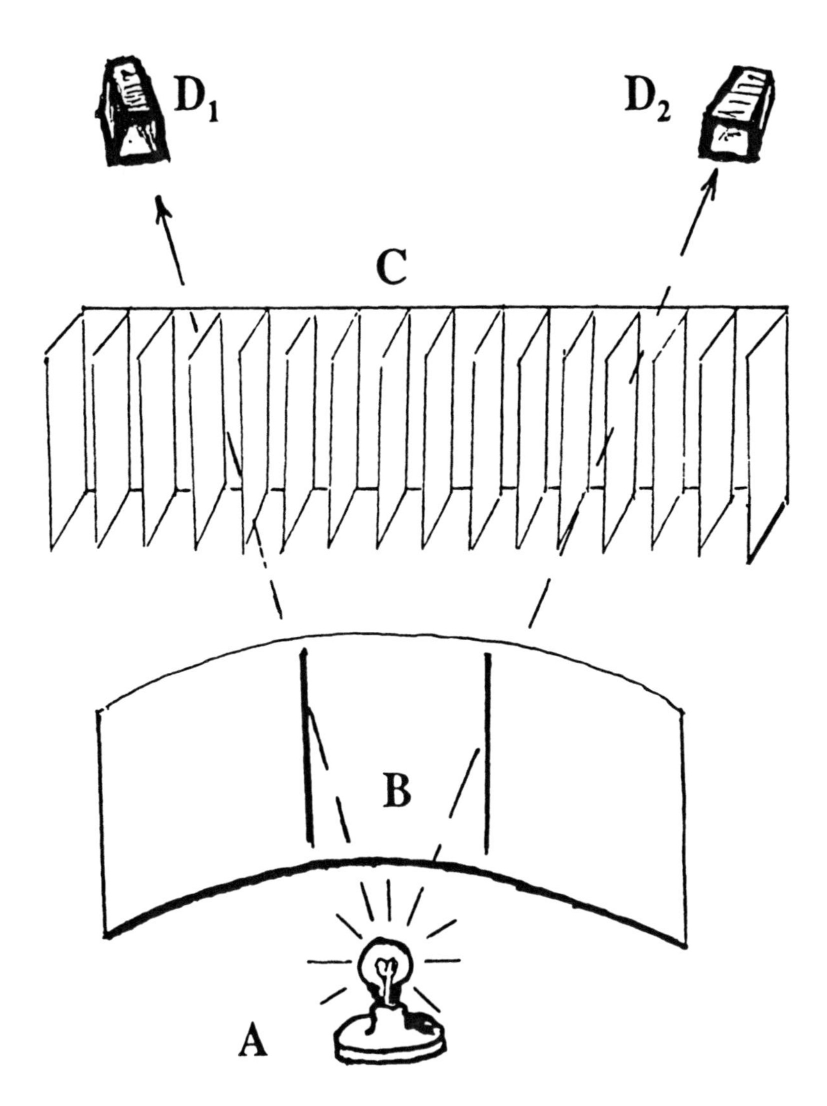

Figure 3. The Delayed-Choice Experiment

experiment is known as the "delayed-choice" experiment because it allows us to decide, after the photon has passed the slits, whether we will observe it behaving as a particle or a wave, passing through one of the slits, or both of them. Like the Aspect experiment, this experiment was first proposed as a thought experiment. But it was actually performed in the mid- eighties by groups working independently at universities in Maryland and Munich.

The results show that, in fact you can wait to make the decision until after the photon should have passed the slits and cause the desired wave or particle behavior any time before the photon reaches the screen. This, of course, agrees with the Copenhagen interpretation: no phenomenon until it registers, either on the screen, or in a counter.

It has been suggested that this experiment violates the arrow of time, allowing us to alter the history of the photon after the fact. Applying this thinking on the cosmological scale, some physicists consider the act of observation, or more specifically, the act of registering the phenomena of light to be an "elementary act of creation". In the case of a photon originating billions of years ago, at the edge of the known universe, traversing the intervening light years of space, and happening to pass through the gravitational lens of a galaxy between there and here, we may choose to receive it in a photon collector, indicating that it travelled on one side of the galactic lens, or we can capture it on a photographic plate as part of an interference pattern, indicating that it had to have passed on both sides of the intervening galaxy. Based on this kind of reasoning, John A. Wheeler, on page 126 of At Home in the Universe, American Institute of Physics, 1994, concludes that:

> It is wrong to think of the past as "already existing" in all detail. The "past" is theory. The past has no existence except that it is recorded in the present...What we have a right to say of past space time, and past events, is decided by choices... made in the near past and now. The phenomena called into being reach backward in time...even to the earliest days of the universe. ...Useful as it is under everyday circumstances to say that the world exists 'out there' independent of us, that view can no longer be upheld. There is a strange sense in which this is a "participatory universe."

We will come back to this point later to discuss just how and in what way reality is participatory, but if we examine the delayed-choice reasoning a little more closely, we find that a subtle confusion is involved. When the photographic-plate venetian blind is open, the photon appears to behave in strict accordance with classical local reality: travelling like a little missile, from its source, through a single slit to register in a collector. When the blind is closed, the photon's flight is influenced by both the slits and some sort of nonlocality is implied. But in fact, Bohr's answer to EPR has again been ignored. His answer was that particles do not exist until an observation is made. Perhaps a better way to say it in the context of the two-slit and delayed-choice experiments is: A photon does not exist locally until it registers in some tangible way, as it does on the blind, or in a collector. Before that, the only meaningful description of the photon involves numerous potential states related to the entire experimental set up. And the localized photon behavior that will become actual depends upon the choice taken by the observer.

Can the photon's past be altered by a choice made just before it might register on the screen? Not if it doesn't yet have a past. If the present is the only time that exists, is it possible that we may be creating the illusion of a past that never really existed by the way we make observations now? This raises some interesting questions about the validity of past histories of the universe that have been pieced together by scientists from observations in the present. Could it be that the big bang and other cosmological, and even geological sequence details, are illusions created by the way we ask our questions, set up our experiments, and make our observations?

Returning to the two slit experiment, we have to ask: Why do the impact points of the individual photons build up an interference pattern on the photographic screen, even when they are released one at a time, no matter how much time elapses between individual impacts? It might be argued that we can preserve local reality if we think of the photons in probabilistic and statistical terms. Assume that individual photons leave the source randomly. They may pass through one of the slits, or strike the opaque material on either side of, or between the slits. Of those that go through one slit or the other, it can be argued that each follows its own specific path through local space, but contributes to the bands of the "interference" pattern because the probability of impact is greater in those areas.

Warming up to this probabilistic approach, we may be tempted to try to apply it to the EPR paradox as well. EPR, of course, refers to the fate of two specific individual particles created in a specific event, while Heisenberg's uncertainty principle is based on probability theory and therefore deals with statistical measurements. Thus it seems that the whole paradox may be explained away as a misunderstanding. Perhaps Einstein and Bohr were simply speaking different languages.

As promising as it sounds, this reasoning fails upon closer examination. Statistical methods are based on probability theory which was developed specifically for dealing with data that exhibit variations due to cause or causes unknown. The danger in applying probability and statistical methods when attempting to discover the nature of reality, lies in forgetting the origin and scope of application of the methods and inadvertently accepting the idea that the real world functions like a statistical model. When statistical methods are invoked, the assumption is made that the causes underlying the observed phenomena are either unknown or unknowable. Thus the investigator is relieved of the task of probing for causes, and the circumstances from which the data derive are forever sealed in a black box. Statistical methods only reveal patterns of variation and distribution within the data, not the details of the process or processes that produced the data.

If we accept a statistical explanation of the dark and light banded pattern in the two slit experiment, we see no need to ask how or where the pattern originates and may conclude that it is sufficient to note that it is a function of the specific experimental set up that is chosen. Photons passing through one slit create one pattern, two slits, another, and so on. Under these assumptions, the bands are not necessarily interference patterns, but simply the building up of photon impacts on the screen according to a probability distribution. But why does this distribution mimic the interference pattern of oscillating waves? With two slits, there should be only two or three bands on the screen. The two directly behind the slits would be the brightest, with a possible third where the two overlap. But in fact, the photons build up many bands, with the brightest occurring in the middle. This pattern is identical to a wave interference pattern.

The attempt to resolve the EPR paradox and explain the two-slit and delayed-choice experiments by application of probability and statistical methods fails because of at least two fatal flaws:

(1) Probability and statistical methods are designed to deal with data when the underlying causes producing the data are unavailable. Statistical moments such as the mean, standard deviation, and variance are of little use in attempting to explain the causes underlying the data resulting from an experiment.
(2) As a result of the Aspect experiment, we know now that attempts to explain the EPR paradox and the two-slit experiments that include only particles, paths, and configuration of experimental equipment, are ignoring the Copenhagen interpretation and the involvement of the observer.

The Copenhagen interpretation and the results of the Aspect experiment force us to return to the question of how consciousness participates in quantum reality. Clearly, observation is meaningless without a conscious observer. We have pointed out the fact that the a priori assumption that consciousness has no direct involvement in quantum phenomena is based upon appearances which are misleading and/or incorrect. Is there a way to investigate the opposite hypothesis: the possibility that consciousness is directly involved in quantum phenomena? As incredible as it may seem, we will see that there is a way, not only to investigate this possibility, but to prove that all the structures we perceive in physical reality are results of the functioning of two different forms of consciousness.

We can start with the fact that the Copenhagen interpretation has been validated by the Aspect experiment. To understand what this really means we must carefully examine the process of observation. Consider the observation of elementary particles in a controlled laboratory experiment. The EPR paradox, the logic of Bell's theorem, and the conclusive results of the Aspect experiment force us to accept the fact that, in such a laboratory experiment, elementary particles are not localized objects separate from the apparatus used to detect them. Therefore, it is not the particles that are measured and observed after they register on the apparatus, it is their effects. And it is not the particles that appear to be separate from the apparatus and the observer, it is their observable effects, and only their observable effects.

Now we know that the apparatus, in fact, the laboratory and everything in it, are all made up of elementary particles which the Copenhagen interpretation tells us also do not exist as separate objects. What we see and interpret as objects are their effects, registering on the receptors of our sense organs by means of the effects of yet other elementary particles, like photons, in the case of the organs of sight. And, of course, the sense organs are made up of the same kinds of elementary particles-- which, in turn, we must remind ourselves, did not exist until they registered on some receptor (see Figure 4). The effects of photons or other particles on the sense organs are relayed to other organic complexes making up the nervous system (ganglia, synapses, etc.) which relay the effects to the brain of the observer. But we can't stop there. The brain is also made up of elementary particles, which do not exist until they have impacted some receptor in some measurable way, but impact upon what? What are the final receptors?

It is here, at the interface of consciousness and matter, that the assumptions of scientific materialism truly break down. Matter and energy only become objective quanta when they register as effects. We know, by virtue of the Copenhagen interpretation, proved by Bell's theorem and the Aspect experiment, that none of the physical quanta moving in this chain from object to conscious observer can exist or exhibit paths, orbits, or any material existence, until they register on the next receptor. And what is the final receptor? It can't be composed of quanta of physical particles and waves because the question would always be: once activated by the effects of the incoming information-bearing quanta, upon what will these quanta register in order to come out of the spectrum of possible states? The chain of registering quanta would have to continue endlessly, constituting an infinite descent. Since a quantum, although very small, has a finite, measurable size, this is impossible.

The inescapable conclusion is that the final receptor completing an observation must be something real, existing beyond physical quanta, capable of receiving and properly organizing the distinct impressions of form and structure relayed by the effects of the quanta of matter and energy that we perceive as existing in an objective physical world. Since the interpretation of form and structure as information with meaning is a function of the conscious

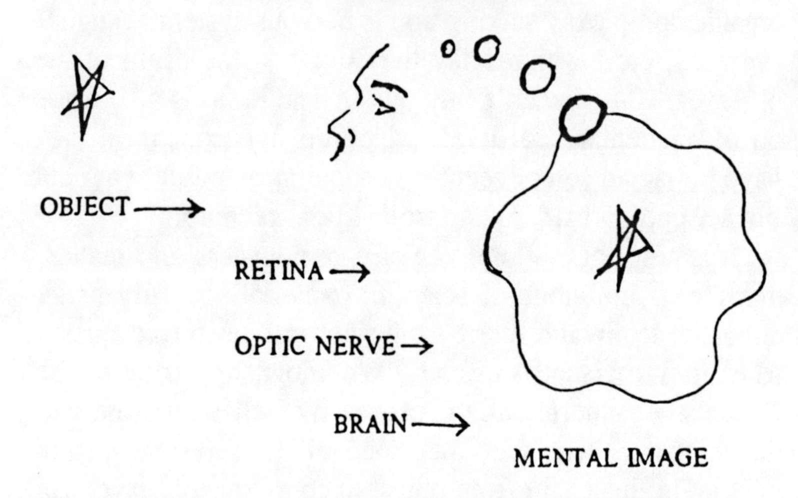

QUANTUM SCALE

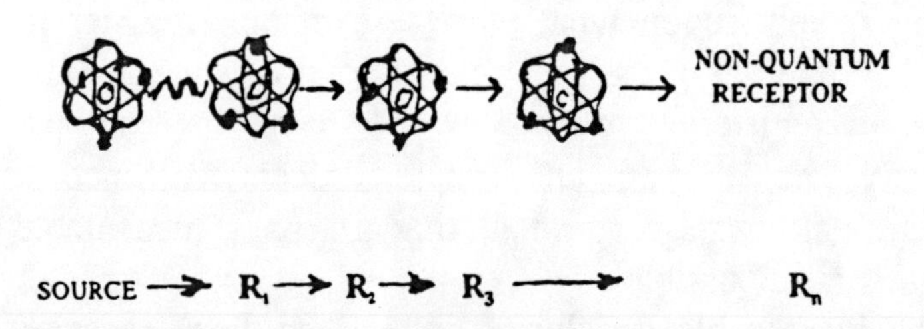

Figure 4. Chain of Receptors

observer, this something must be closely related to, if not the very substance of, consciousness itself. It follows that the final receptor or conscious component in a conscious observer is therefore real and substantial, but non-physical.

Since "non-physical" is a term that may be easily misinterpreted, we must be clear about the meaning intended. In this context it simply means: not composed of quanta of matter and/or energy. It does not imply that this component is not just as real and substantial as the quanta and their effects in the physical universe. In order to avoid, as long as we can, inventing new terms, we will call this conscious substance individualized consciousness. This is, of course, a specialized use of the word consciousness. Throughout the remainder of this book, the term individualized consciousness will be used to refer to the real substance at the end of the chain of receptors associated with a conscious observer. Thus observation is an activity performed by individualized consciousness, and awareness is a result of this activity.

This argument proving that consciousness is ultimately necessary for the formation of any physical object is formalized in Appendix C. The proof, which is technically beyond the scope of this discussion, is based on the mathematical logic of the Calculus of Distinctions and the concept of the conscious drawing of distinctions, which are discussed in detail in Chapter 6 and Appendix D.

The logical procedure of infinite descent may also be applied to the entire universe: If we imagine time as we seem to experience it, running backward to a point before any form of life existed, no observation of the type discussed above, involving sentient organic beings, could be made. Could all the forms and structures that we perceive in the universe, including solids, liquids, and gases, mountains, planets, and galaxies, all made up of elementary particles, with the exception of those of conscious life forms, have existed at that time? It would seem so. But all physical forms are effects of elementary particles, and if we continue to roll time back toward the big bang, we have to ask: what was the receptor of the first particle or particles?

We know now that without a receptor, no form or structure could emerge. Without receptors to record their effects, all of the elementary particles would remain in the nonobjective spectrum of possible states represented by Schrödinger's probability wave function, expanding forever at the speed of light. Could there have been a non-physical receptor? We've just shown, in the case of the conscious observer, not only that there can be, but that there has to be a real, but non-physical receptor, a conscious substance beyond quanta. And the fact that the physical world does exist at this moment proves that a non-physical receptor existed prior to the creation of the physical universe.

You might argue that all the receptors necessary for a given observation, the quanta of the atoms making up the equipment, the scientist's eyes and brain, and everything else in the chain of observation, existed prior to setting up the apparatus to observe the elementary particles in a two-slit or EPR/Aspect type experiment, or any other experiment. But no matter how long they have existed, as elementary quanta they had to have had receptors. If those receptors were physical, i.e., composed of other quanta, they also had to have had receptors. Clearly, we cannot avoid infinite descent without a non-quantum receptor.

While trying to reduce physical reality to its most basic units: quanta of matter and energy, science has proved, through Bell's theorem and the Aspect experiment, that reality is nonlocal. What can the nature of this nonlocality possibly be? It is very curious that while we describe our perceptions of reality in terms of objects and observations, the only thing that we ever actually experience directly is our own consciousness. And it takes only a moment's reflection to realize that a primary feature of that experience is nonlocality. Individualized consciousness is not normally restricted to the awareness of one point at a time, but encompasses a region, such as the head, abdomen, hands, or even the entire body. I can choose to think of the top of my head and the bottom of my feet at the same time. In this way, I can experience simultaneous awareness of parts of my anatomy that are at opposite ends of my body. I can also be aware of conditions in different rooms in my home, or of events in two or more distant cities at the same time.

In spite of the quantum nature of the physical world revealed by scientific experiment, we experience a continuous field of awareness that may be focused on a specific region, or expanded to encompass the entire body. We are also aware, through the interpretation of sense data, of regions beyond the boundaries of our physical bodies.

Since the only thing we know that has this feature is our own consciousness, it seems likely that the nonlocality that pervades the universe may be a form of consciousness also. Thus the most reasonable hypothesis is that the non-quantum receptor that preceded the creation of the physical universe is the same nonlocal consciousness that pervades the universe. In order to discuss this hypothesis, we need terminology that will consistently describe this form of consciousness. Because of its all-pervasive nature as the background of all phenomena, it seems appropriate to call it primary consciousness.

John von Neumann discusses the necessity of something "extra-physical" in Section VI of Die Grundlagen. He explains, using the example of temperature, how calculations, and in particular, quantum mechanical representations, can describe the processes of measurement and observation. He discusses details of the measurement, including the physical dynamics of heat expansion and the thermometer, the reflection of light from the mercury column into the eye of the observer, the formation of an image on the retina, etc., right into the brain of the observer, concluding with:

> ...and then in the end [we must] say: these chemical changes of his brain cells are perceived by the observer. But in any case, no matter how far we calculate -- to the mercury vessel, to the scale of the thermometer, to the retina, or into the brain, at some time we must say: and this is perceived by the observer. That is, we must divide the world into two parts, the one being the observed system, the other the observer. In the former, we can follow up all physical processes (in principle at least) arbitrarily precisely. In the latter, this is meaningless. The boundary between the two is arbitrary to a very large extent...
>
> Now quantum mechanics describes the events which occur in the observed portions of our world, so long as they do not interact with the observing portion, ... However, the danger lies in the fact that the principle of the psycho-parallelism is violated, so long as it is not shown that the boundary between the observed system and the observer can be displaced arbitrarily in the sense given above.

A suggestion of infinite descent can be seen in these passages and, indeed, Wigner and others followed this lead. This is all the more impressive because at the time it was written, von Neumann did not have the benefit of Bell's theorem and the Aspect experiment. There is no evidence that he wanted to believe in any sort of non-physical reality. His conclusions are based on logic, not personal bias. However, in spite of his brilliance and insight, von Neumann did not make the transition to understanding consciousness as an essentially non-quantum substance, and we will see that his pronouncement of the meaninglessness of describing processes in the consciousness of the observer is unwarranted, but understandable in the context of scientific materialism.

At present, however, we are entering a new era of science and the evidence that we have developed in this chapter logically forces us to reverse the basic assumptions of scientific materialism. If the non-quantum receptors are forms of consciousness, the assumptions that the physical universe could exist as it does without the presence of consciousness, and that consciousness, as it appears in living organisms, is a chance outgrowth of the evolution of the material universe are simply wrong. The premise of transcendental physics, supported by Bell's theorem, the Aspect experiment and proved by the logic of infinite descent, is that a primary form of consciousness had to exist prior to the manifestation of any physical object.

The form and structure revealed by the first quanta cannot have originated from a physical cause, since none existed. The physical evolution of the universe is therefore the unfolding of the innate structure of non-quantum primary consciousness. As individual conscious entities, we participate in this unfolding by making ever-increasing numbers of distinct, individual observations.

We might try to avoid this conclusion by arguing that it is possible that, although elementary particles cannot exist without receptors, we can perceive the forms that we do while the elementary particles exist only as probabilities described by Schrödinger's wave equations. But the research of Rutherford, Bohr, and others has shown that the structures of the physical universe, from the gases burning in distant stars, to the crystalline molecules of metals and stone, and the organic molecules making up a brain, all depend upon the atomic structure of the elements of which they are composed. Furthermore, this atomic structure depends upon the existence of elementary quanta, and the existence of quanta ultimately depends upon reception in consciousness. As it is illogical to assume that effects can exist without cause, we have to conclude that observable reality depends upon quanta, and that non-quantum consciousness, pervasive, and individualized, are necessary ingredients of reality, participating in the processes that cause the universe to exist as it does.

The discovery that consciousness is necessary for the existence of a physical universe has far reaching implications. The time-honored, most basic assumptions of scientific materialism are wrong! The physical world we perceive and study as scientists cannot be considered to be separate from the observer, as Einstein, and just about everyone else thought. Not only that, the twin-particle effect documented by Aspect demonstrates that a basic feature of reality is the direct connection of all parts of the universe, however distant.

Physicist F. David Peat, in his book Einstein's Moon, Contemporary Books, 1990, said on page 121:

> Scientists agree that the many different experiments employed to test John Bell's predictions come down

> firmly in favor of orthodox quantum theory. Einstein's position is now untenable --- local reality cannot be retained.

Bell used the term nonlocal to describe the reality implied by what is now the orthodox Copenhagen interpretation of quantum theory. This term has no well-defined meaning in the current paradigm. Continuing with Peat's comments, on page 124 of the reference cited above:

> Reality can no longer be restricted to a purely "local" meaning, for the nonlocal implications of Bell's theorem show that what happens in one region of space is correlated with other distant regions in the universe. ...our concepts of space and time may need to be transcended. The implications of Bell's theorem may well demand an even greater revolution in thought than what had first been expected from relativity and quantum theory.

And Roger Penrose, Professor of Mathematics at Oxford University, in his book Shadows of the Mind, subtitled A Search for the Missing Science of Consciousness, says on page 419:

> If Einstein's general relativity has shown how our very notions of space and time have had to shift, and become more mysterious and mathematical, then it is quantum mechanics that has shown, to an even greater extent, how our concept of matter has suffered a similar fate. Not just matter, but our very notions of actuality have become profoundly disturbed.

Some scientists are convinced of the need for a new paradigm, but no one seems to know how to produce one. The idea of a reality "out there" or, as Einstein said, " an external world independent of the perceiving subject" is too much a part of our world view and our science to be discarded easily, even when hard evidence and logic say that we should. Those believing in an out-moded paradigm, will try to explain away any new fact or observation that does not fit that paradigm. If something can't be explained in terms of the accepted view, they will categorize it as an anomaly and simply ignore it. That's what has been happening since 1982, for the most part. Physics and materialism have been synonymous for most of the history of science. Not many physicists want to throw out the most basic assumption of materialism--but the evidence is very compelling, and New basic assumptions are what make new paradigms.

Non-quantum consciousness is an integral part of reality, and thus it is clear that an effective new paradigm cannot be based on the assumptions of scientific materialism. It must incorporate the reality of non-quantum receptors. We must dispel the wide-spread confusion of objectivity with the current view of physical reality, and deal with new information which indicates that reality exists as a spectrum of substance, ranging from gross matter to energy, to more and more subtle forms, and finally, to the non-physical substance beyond the quantum. The only word we have in the current scientific lexicon that even approximates an appropriate description of this subtle substance is the word consciousness.

A growing number of scientists, like Roger Penrose, appear to be on the right track, at least recognizing the direction we have to go. On pages 388 and 420 of the work referenced above, he says:

> There are reasons for being suspicious of our physical notions of time, not just in relation to consciousness, but in relation to physics itself, when quantum nonlocality and counterfactuality are involved... we are presented with a profound puzzle. ...
>
> These are very deep issues, and we are yet very far from explanations. I would argue that no clear answers will come forward unless the interrelating features of all these worlds [physical, mental, and mathematical] are seen to come into play. No one of these issues will be resolved in isolation from the others...No doubt there are not really three worlds but one, the true nature of which we do not even glimpse at present.

The one-world argument put forth here by Penrose is an important concept in the development of transcendental physics. There is only one reality, and it includes everything from non-quantum receptors to the smallest quantum, to the largest galaxy in the universe. A successful new paradigm must integrate all the valid mental and mathematical models of reality with each other and with physical reality. This cannot be done with the limiting assumptions of scientific materialism.

With proof that materialism is an incomplete and inadequate theory, science can no longer use it as a valid basis for understanding reality. Ironically, it is not a thought experiment or theoretical consideration that has brought about the downfall of materialism, it is a meticulous empirical experiment. Scientific materialism has literally been hoisted on its own petard! And the blow is fatal. The assumption that physical reality exists independent of the conscious observer, is simply incorrect. Still, because of our ingrained belief in this assumption, the death of materialism is not yet general knowledge. We have been enthralled by the power of material technology, and temporarily seduced by the illusions of materialism, but eventually the truth will prevail.

As reasonable as the assumption of the independent existence of the world of physical effects seems, there is no real evidence to support it. We can no longer pretend that the apparatus needed to observe elementary or sub-atomic particles could have evolved by themselves; much less the eye of the observer. The logic of Bell's theorem, the evidence of the Aspect experiment, and the proof of the existence of non-quantum reality by infinite descent indicate that the particles and waves of physical reality remain in a state of multiple possibilities until an observation or measurement is made. Reality is simply not the reality we perceive until we perceive it. If our means of perception were constructed in a way that would select a different set of effects out of the spectrum of possible states, a different reality would appear. In the two-slit experiments, for example, light can be caused to manifest either as waves or particles, depending upon a choice made by the observer.

In the next chapter, we will have a look at the brain, the receiver of information from the external world in the form of physical signals such as photons, from the point of view of quantum mechanics.

CHAPTER 3. MIND AND MATTER

A Materialist's Dilemma

Many of us who attended the April, 1996 conference *Toward a Science of Consciousness* in Tucson Arizona, were fascinated by the debate that raged between those who believe that consciousness either has been explained, or will soon be, by our detailed knowledge of neurology and the physics and chemistry of the processes of observation and perception, and those who think that there is something more to consciousness than the interaction of matter and energy in time and space, something that transcends the known elements and parameters of material science as it exists today.

The great mathematician, scientist, philosopher, physician, linguist, etc., etc., Gottfried Wilhelm von Leibnitz did not believe that mind, or consciousness, could ever be explained completely as a function of physical structure, however complex that structure might be. He imagined the physical structures of a human being duplicated exactly, but on a much larger scale, so that the inner workings could be studied like the levers and gears of a windmill. Studying the details of the brain, he reasoned, would increase our knowledge of how the brain functions within the body, but nothing would be learned from such an exercise concerning the nature of mind itself.

An up-dated version of this argument, which I believe sharpens the point a bit further, was presented by A. C. Elitzur of Tel-Aviv University at the '96 Tucson Conference. Dr. Elitzur proposed a challenge to the quintessential materialist in the following manner: If every aspect of your consciousness, every memory, thought, and characteristic can be accounted for by the quanta of matter and energy making up your brain and body, then, since the type, number, and size of quanta are finite, at some point in the future, we will be able to assemble an exact duplicate of you. As a materialist, it should not concern you, if, after you are duplicated, your original body is destroyed, since everything that makes up your consciousness would still exist in the duplicate. Dr. Elitzur didn't think that even the most adamant materialist would say without hesitation: "O.K., go ahead and kill me, I'll go home in the new body." Even a materialist might pause and wonder if there could be something wrong with this setup. If the total configuration of matter and energy that makes up your physical being is no longer unique, could there be *something* else, other than matter and energy, that makes your consciousness different from that of any other physical being, even if that being is physically identical to you in every way?

Dr. Elitzur's materialist's challenge brings up some interesting questions: Once the duplicate has been assembled, will your consciousness inhabit both bodies at the same time? Will you know which is your original body and which is the clone? If, as the materialist insists, everything must be explained in terms of matter and energy, the answer to the first question is yes, since there will be no physical difference, and thus nothing to allow a difference in consciousness to be drawn; and the answer to the second question is no, for exactly the same reason.

Consciousness and Quantum Mechanics

Planck, Einstein, Bohr, Schrödinger, Heisenberg, et. al., did not set out to explain consciousness. They were only trying to explain sub-atomic phenomena. As it turned out, this couldn't be done without including the consciousness of the observer in the account. As a result, the most significant advance toward a science of consciousness since Leibnitz's time was, and still is, the

Copenhagen interpretation of quantum mechanics. Those who don't think so, to paraphrase Bohr, simply haven't understood quantum mechanics. What makes quantum mechanics a significant advance toward a scientific understanding of consciousness is not that it provides a one-to-one correspondence between every aspect of consciousness and some configuration of quanta of matter and energy, but its revelation that various aspects of consciousness like randomness, chance, nonlocality, and complementarity, are inherent in the very fabric of physical reality.

Can consciousness be explained in terms of matter and energy as they are presently defined? Even if it cannot, this doesn't mean that a scientific understanding of consciousness is not possible. There may be more subtle forms of the substance of reality that we haven't yet discovered. And the idea that the explanation of consciousness may lie outside the parameters of matter, energy, time and space as currently defined, doesn't mean that there isn't anything to be gained by investigating the relation of the functioning of consciousness to the physical structure of the central nervous systems of sentient organisms. To the contrary, such an investigation, especially using the new concepts of quantum mechanics demonstrated by Bell's theorem, the Aspect experiment, and the delayed-choice experiments discussed in the last chapter, may well lead to a greater understanding of what consciousness is and how it fits into the universe. Such an understanding may lead to a reversal of the traditional scientific attitude of excluding consciousness from the objective study of the realities we experience. Henry P. Stapp, of the University of California at Berkeley, ably discusses the need for, resistance to, and inevitability (largely because of the results of Bell's theorem and the

Aspect experiment) of this reversal in Chapter 11 of his book Mind, Matter, and Quantum Mechanics, Springer-Verlag, 1993. He closes this chapter with the following words:

[If] the nonclassical mathematical regularities identified by quantum theory are accepted as characteristics of the world itself, a world whose primal stuff is therefore essentially idealike, and if, moreover, these mathematical properties account in a natural and understandable way for the classical characteristics of our conscious perceptions, as they seem to do, then we appear to have found in quantum theory the foundation for a science that may be able to deal successfully, in a mathematically and logically coherent way with the full range of scientific thought, from atomic physics, to biology, to cosmology, including also the area that had been so mysterious within the framework of classical physics, namely the connection between processes in human brains and the stream of human conscious experience.

A growing number of physicists see quantum theory as the doorway to a new science. We've discussed some of the reasons for this in the previous chapters. Now, let's have a look at how quantum theory may be able to describe the mind- body connection.

Quantum Theory and Brain Dynamics

So the brain is made up of quanta of matter and energy. Isn't everything? Why are quantum physicists becoming so involved in the study of consciousness? Can quantum theory shed some light on the experiences of consciousness? First, theoretical physicists tend to be idealists. Many of them are searching for the Holy Grail of science: the key to the ultimate nature of reality, an idea or discovery that will lead to an all-encompassing theory. The deep metaphysical leanings of some physicists, like Einstein, lead them to dream of a unifying theory, i.e., a "theory of everything". Among other things, a Grand Unifying Theory must explain consciousness. Will quantum theory, if it can be integrated with relativity, lead to this goal? To understand why quantum physics truly is an important advancement that may help develop a science of consciousness, we need to know how quantum mechanics relates to brain dynamics.

We shall not try to present a text-book explanation of quantum theory here, or a detailed discussion of what is known of brain dynamics at the quantum level. This has been done very well by several authors, notably by Henry Stapp (Mind, Matter, and Quantum Mechanics, Springer Verlag, 1993), Roger Penrose (Shadows of the Mind, Oxford University Press, 1994), and Mari Jibu & Kunio Yasue (Quantum Brain Dynamics and Consciousness, John Benjamins Publishing, 1995). We will attempt to give a brief overview of quantum brain dynamics, just enough to allow the reader to understand the basics and follow the arguments and conclusions presented later in this book. Someone may object that by doing this, we might be overlooking important details that could put our conclusions in question; but the object here is to avoid becoming entangled in an excessively abstract discussion of the mathematical details of quantum theory and unnecessarily bogged down in relating biological microstructure details to the functioning of consciousness. Such relationships are controversial at any rate. Also, as we shall see, a new mathematical language without the numerical limitations of matrix algebra is needed to deal with the more subtle questions of how consciousness interacts with matter and energy.

The purpose of our discussion of quantum-level brain

dynamics is to attempt to identify phenomena occurring within the physical structure of the brains of sentient beings that correspond to conscious experiences such as sensory processing, thought, memory, and the internal perception of images. The brain cells that make up neurons, with their dendrites, axons,and synapses are all composed of molecules, composed of atoms, composed of hadrons and leptons, composed of electrons, photons, gluons, and quarks. And the physical transfer of energy within the brain that gives rise to self-awareness, thought, and other conscious activity is believed to be accomplished primarily by the exchange of photons and electrons with the help of gluons. Photons and electrons are elementary particles that may pass from the external world of objects to the internal world of neural networks by the way of the sense organs. They also convey from the external world, across a great many neural synapses, the information necessary for the internal visualization of scenes representing the external world.

As we discovered in the last chapter, elementary particles are not simple physical objects like tiny baseballs. They have both wave and particle characteristics, and perhaps even other characteristics that have not yet been identified. <u>And</u>, as the Copenhagen interpretation and the results of the Aspect experiment tell us, they do not exist as particles or waves until they interact with some other physical object or structure. En route from a source to a receptor, they exist as only as *a range of possibilities* that is described by the Schrödinger wave equation.

When a quantum, say a photon, impacts on a receptor and its effect is observed or measured, we say that the multi-state wave function has collapsed, allowing the photon to register its effect as a single space-time event. Exactly how and why this collapse occurs, and whether consciousness has anything to do with it, is a question of intense debate and research. Most physicists, at least those who want to maintain a materialistic point of view, are very hesitant to consider the possibility that the act of conscious observation could cause the collapse. We do know, however, since it has been demonstrated by the double-slit experiments described in the last chapter, that choices made by a conscious observer can determine the form of matter or energy, e.g., wave or particle, that is actually observed.

How does a wave-function range of probabilities that describes a quantum of energy moving in the neural network of the brain become a single measurable effect? Before we attempt to find the answer to this question, let's trace the development of quantum theory in a little more detail: In 1905, Einstein published a paper describing the theoretical interaction of photons and electrons as quanta. This idea was based on the experimental work of Max Planck and the matter-wave theory of Louis de Broglie. With the discovery of the Compton effect in 1923, the quantized interaction of photons and electrons was confirmed. Until that time, light was considered to be an electromagnetic field phenomenon, best described by Maxwell's wave equations. In recent years, physicists like Feynman, Schwinger, Mills, Higgs, Tomonaga, Weinberg, Umezawa, and many others, have been hard at work trying to remold electromagnetic field theory into a quantum field theory. While there still are some details to be worked out, a viable theory, called quantum electrodynamics, or QED, which describes the physical phenomena involving electrons and photons has been developed. In general, this quantum field theory provides a logical framework for describing the interaction of quanta of matter and energy in the brain, i.e., quantum brain dynamics. These quantum processes underlie biological processes, and are assumed to underlie psychological functions.

As is the case in any quantum system, the elementary quantum components of the brain are related in complex ways that can be described by quantum field theory. A key feature of quantum electrodynamics is the quantum coherence of billions of individual quanta affected by or participating in a field, that can occur as waves or pulses of energy moving through the mass of quanta making up the field. This may be related to the phenomenon of nonlocality discovered through the EPR paradox, Bell's theorem, and the Aspect-type experiment, and it is a bit easier to conceptualize. We can compare it to the way a school of fish or flock of birds moves. In such a case, a multitude of individuals act as if they are all controlled by forces or impulses felt throughout the entire mass or swarm of individuals. Such coherent patterns can flash through the brain mass, and they can change from one pattern to another almost instantaneously. These patterns of energy movement in the form of electrons and/or photons can be said to correspond to thoughts, and complex superposed groups of such patterns may correspond to states of mind.

Now, let's consider how these flashing coherent patterns arise: If we consider the brain, central nervous system, and sense organs of a sentient individual as dynamic, interacting quantum systems, then when quanta of energy impact these systems from outside, through the sense organs, i.e., eyes, ears, skin, tongue, etc., if the incoming impulse is above the threshold level of the sensitivity of the receiving quantum systems, the resulting patterns moving through the interacting quantum systems of the individual (or observer) can be understood as automatic quantum field responses. If the physical systems of two observers are similar enough, as is the case with human beings, the automatic responses will be similar enough that they will usually agree, in general, about what they have seen, heard, tasted, etc. Differences in the interpretations of new experiences will come up, however, because of the internal comparison of the new patterns, arising from the incoming impulses, with patterns stored in individual memory from past experience. The details of these remembered patterns will be unique for each individual, because of the differences between the actual past experiences of individuals, and because of changes that may occur in the memory patterns of different brains over time.

Quantum Mechanics and Free Will

So far, quantum field theory appears to be admirably suited to describing and explaining the interaction of mind and matter and the functions of consciousness completely in physical terms. However, two problems arise that require us to go beyond the bounds of physical quantum theory. First, quantum electrodynamic field theory does not satisfactorily explain memory: How can part of an interactive quantum system maintain a long-term standing pattern? The answer that they persist as stable low-energy, or vacuum-state quantum systems does not solve the problem. A vacuum state for the entire quantum dynamic brain system that is modified by incoming information-bearing quanta to reconfigure into a new vacuum state offers a way to understanding the process of learning, but does not explain how whole sequences of memories, that should have long since been replaced by similar new experiences, may be recalled in detail, like motion picture re-runs. Second, if every aspect of consciousness has a physical quantum mechanical explanation, where does volition come in? Is the apparent freedom of sentient beings to make decisions and choices an illusion? The answer that quantum uncertainty allows a certain amount of freedom for the individual to exercise choice, does present a window in the physical theory for conscious decision to affect the quantum mechanical system, but does not explain how this occurs.

If we are, in fact, able to make decisions and initiate actions that control the movement of our heads and limbs, direct our thoughts and speech, and manage at least some aspects of our lives, as we believe we are able to do, then some of the energy patterns corresponding to thoughts must originate from within. If the energy initiating quantum processes corresponding to mental functions were to come only from outside, we would be automatons, dependent upon external stimuli for our every thought. This however, is obviously not the case. Energy derived from sunlight, food, water, and oxygen are available internally. Exactly how this energy is brought into quantum brain processes, and used purposefully by consciousness, is another question, one of biology and neurophysiology.

The Experience of Consciousness

In an effort to maintain objectivity, scientists have pretended that we are observers looking at everything from the outside. There is no law, however, that says that this is the only way to assure objectivity in a scientific investigation. In fact, to insist on this stance when investigating consciousness, is ridiculous. The fact that we are conscious beings means that we can conduct objective studies and report on consciousness from the inside. Furthermore, when investigating the interaction of consciousness with the physical world, we have the advantage of being able to look at the situation from both sides. Trying to understand how consciousness works in the brain while limiting our study entirely to the physical aspects of the brain, is like trying to taste a peach through a microscope.

From the inside, our experience of consciousness arises in the form of a sense of being that manifests as the awareness of self, as opposed to the awareness of that which is other than self, i.e., the rest of the world. This self awareness seems to originate in the brain and central nervous system, but extends in a nonlocal fashion to include everything contained within the skin covering the body. We also have experiences that testify to the fact that not all of the physical parts of the body are necessary to the experience of self awareness. Our experience of self awareness is not diminished by the loss of a limb, eye or ear. Under the influence of hypnosis, sleep, in deep concentration, or even just in a state of focused attention, a person sometimes has no awareness of the position, condition, or existence of parts of the body, arms, legs, etc., or even of the entire physical body.

Such experiences make us wonder whether individual consciousness might be able to exist outside of, or without the benefit of, a physical body. The religions of the world, of course, are based on the premise that consciousness is able to exist outside the body, and in fact does exist after the death and disintegration of the physical body. As long a science is conducted from outside, this is a question of belief or faith, or perhaps anecdotal experience. But scientists are conscious and therefore are able experiment and make observations from within consciousness. It is only the assumption that physical reality exists totally independent of consciousness that has prevented serious scientific investigation of consciousness. With the development of a new system of mathematical logic which includes consciousness, science can transcend the limits of the materialistic point of view and investigate such questions as the continuation of conscious existence beyond physical death.

Many people have reported experiences that support the idea that consciousness is something more than an abstract function of quantum mechanical systems. In the last few years, reports of near-death, out-of-the-body experiences have become almost commonplace, and there have been numerous cases of past-life memory documented in various locations around the world, over the past 150 years. It is not hard for a scientist to dismiss such reports as hallucination, imagination, or wishful thinking, as long as the scientist has no way to investigate them directly. It is against the true spirit of science, however, to accept or reject hypotheses on the basis of unsubstantiated theory or belief. We must develop a approach, grounded in logic, and supported by a mathematical descriptive language, that will permit the objective investigation of phenomena related to the question of whether or not consciousness may exist outside the physical body.

Can Consciousness Be Explained in terms of Matter?

Only the most closed-minded materialist could categorically deny the existence of the whole realm of consciousness, or that there would be doorways into it, for the very fact that we can focus on the external world, as classical science has, by excluding any thought of the internal world, certainly acknowledges the existence of some kind of internal world. The symmetry of experiential reality implies that, if you can face one direction, there has to be an opposite direction.

Can the experience we call consciousness be explained fully in terms of quanta of matter and energy interacting in time and space? If by "explain" we mean that we can find physical correlates for thoughts, then the answer is: probably. But if we mean to explain the experience of conscious awareness, the question is a far more difficult one. Can there be such a thing as thought-free awareness? Meditation practitioners around the world attest to the fact that there is a state of thought-free conscious awareness. If their claims are valid, then the discovery of quantum brain processes corresponding to thoughts and images will not explain consciousness, only some of the surface manifestations of it.

Science seems to have lost its metaphysical roots. Many of today's scientists are agnostic, some even aggressively atheistic. I believe that this is due to a misunderstanding and misinterpretation of the scientific method. Basic assumptions must be questioned any time new knowledge conflicts with the accepted paradigm. Recognition of a non-quantum reality underlying the physical universe need not signal a regression into superstition and blind mystical belief. The great ideas that advanced science have always had a metaphysical basis, and many of the great scientists of the past were mystics at heart, as we will see in the next chapter.

Dark clouds of mist, bright drops of rain:
Forms from the void, and yet again,

Electric dance dispels the night
With sheets of fire, of heat and light.

And warriors charge in from the cold,
To beat their shields like knights of old.

See them rise, hear how they roar,
They seize the chance, just this once more,

To try to squeeze a golden scale
From the smokey dragon's slippery tail!

CHAPTER 4. MATERIALISTS AND MYSTICS

Confronted by the dangers with which the advances of science can, if employed for evil, face him, man has a need for a "supplement of soul" and he must force himself to acquire it promptly before it is too late. It is the duty of those who have the mission of being the spiritual or intellectual guides of humanity to labour to awaken in it this supplement of the soul.

-Prince Louis de Broglie, Physics and Microphysics, Pantheon, New York, 1955

Toward a New Science

The expansion of science beyond the study of matter, energy, time, and space is not a trivial event. It is a significant departure from what has gone before. It is a recognition that any description of reality that is limited to matter, energy, time, and space is necessarily incomplete. And it is the recognition of consciousness as a real substance. Because of the scope of subtle phenomena and abstract concepts that comprise consciousness, it is important to take a moment to put the pursuit of science into the proper perspective. The claim that physical science is incomplete without the inclusion of non-quantum and therefore non-physical reality is not an attack on science or intellectualism. Science will always be incomplete, and the expansion of the application of the scientific method into new realms is totally in keeping with the great tradition of scientists like Kepler, Newton, Einstein, and Bohr. In this chapter we will take a brief look at that tradition.

Science and Pre-conceived Ideas

When I enrolled in college as a physics major in 1955, my first physics instructor, Professor Abernathy, who retired after my Freshman year, was about the same age as Sir Arthur Eddington or Albert Einstein. His lectures were crisp and engaging and his physics labs were fascinating. But there was one thing about him that really surprised me: He didn't believe that man would ever set foot on the moon - much less any other planet.

"There is simply no basis for it." He said. A competent engineer, as well as a physicist, he had done the calculations, based on the best information available, to prove his point. The energy and life-support requirements were too great. Even if the physics might become marginally possible with future technological advances, he reasoned, it wouldn't ever be economically feasible.

"To what purpose?" He asked. The moon is an airless, barren ball of rock, composed of the same elements as the earth. Any mineral ore existing on the moon would be far more accessible, even in the most remote, rugged terrain here on earth.

"For the advancement of science? Extremely impractical!" He declared. "Too nebulous a goal. No one in their right mind would spend billions of dollars trying to accomplish such a difficult, fool-hardy thing with such a vague, improbable payoff."

Was he entirely wrong? After one giant step in 1969, mankind, inspired more by the cold war between the United States and Russia than by scientific fervor, seems to have lost the will to reach for the stars. The man-in-space program has fizzled for lack of sufficient funding.

Perhaps no scientist has ever been 100% right - or 100% wrong. Scientists, like everyone else, are subject to preconceived ideas, unsubstantiated beliefs, and erroneous assumptions. Yet the public perception is that if a scientist has the proper credentials, i.e., degrees from the right schools, consistent endorsement from those within the current paradigm, and in general, fits the preconceived mold of scientist, he is practically infallible. This perception is promoted in our academic and intellectual communities, and is similar to the attitude the average person had toward medical doctors a couple of decades ago. An important ingredient of the current scientist profile, practically a requirement for admission to the ranks of the scientific establishment, includes a materialistic attitude.

To what extent does attitude affect the outcome and conclusions of scientific studies and experiments? Without thinking, most of us would say: Not much, if any. We are conditioned from early childhood to believe in a solid physical reality that exists "out there" objectively, independent of our thoughts and attitudes. And the purpose of science, of course, is to determine the nature of that objective reality. But now, with the knowledge of Bell's theorem and experimental results validating the Copenhagen interpretation, we have to re-think this attitude. Our state of consciousness and the choices we consciously make have been shown to affect the form of the physical reality we observe.

The Search Takes a New Direction

We must remember that the Aspect experiment proves that none of the particles and waves that make up the objects that we see, and even the light rays with which we see them, exist as such before they register in a way that allows the effects to be observed. We know now that, until they register, they have no independent objective form, no pathways, and no history. And now we have proved with the logic of infinite descent that they cannot register at all without the involvement of consciousness.

The limited nonlocality of consciousness in the receptorium of individual awareness in sentient beings is necessary to provide end-point receptors to avoid the contradiction of infinite regression in the finite world of quanta. Furthermore, a primary, non-quantum based form of consciousness had to exist before the first quantum could congeal out of the big bang. Our understanding of the nature of reality and the universe that we perceive has suddenly changed. And this change is a very radical change from the world view still being advanced by the established coterie of scientific materialism.

The Need to Know

Curiosity is one of the most basic characteristics of human nature. We are driven by a need to know. The very young often express this curiosity freely: Why is the sky blue? Where does the sun go at night? Where do babies come from? As we grow older and learn, our questions become more sophisticated. We may even come to question the very meaning of existence. Why are we here? What is the ultimate nature of reality? We find ourselves somehow consciously aware of a vast and mysterious universe that seems to exist outside and independent of what we think of as ourselves. Now we know that this external world is not what it seems. It is not what science has taken it to be for the past two centuries. It is much more.

During the early stages of the development of civilization, we turned to religion for our answers. Religion provided a context for explanation, especially for things we could not investigate directly. In the seventeenth century, science and religion were scarcely separate at all. In the West, natural science, inspired by Greek thought, was a blend of philosophical and religious reasoning with an increasing application of experimental verification. Science was born within the context of religion, and at first, most science was carried forward by monks working in monasteries. But like many other social institutions, organized religion created a self-serving paradigm that increasingly discouraged and repressed man's natural curiosity. When science began to ask questions that religionists found threatening to their dogma, and announce conclusions that were contrary to long-held articles of faith, religionists denounced it as heresy and began to persecute those practitioners of science who were so bold as to cast doubt on religious dogma. Whenever social or political institutions develop to the point of suppressing our innate desire to know and understand, we have to find new means to investigate the unknown, to satisfy our need to know.

The Separation of Science and Religion

When science began to embrace materialism as its primary philosophical basis around the end of the eighteenth century, it did so for some very good reasons. As a budding intellectual institution, science had to divest itself of the liabilities of its roots in alchemy, astrology, and numerology and establish an unassailable power base by applying itself to easily provable propositions. There was a world of useful knowledge waiting to be discovered through simple, straight-forward research, and it gained science nothing to have its practitioners burned at the stake as witches. Unfortunately, the swing to materialism that resulted was a clear case of the proverbial throwing the baby out with the bath water.

In order to avoid the uncertainty of individual prejudice and belief, scientists denounced everything mystical. As a result, science lost its metaphysical grounding and, like government, was also separated from religion. This was especially true in the West. As science began to establish itself under the aegis of materialism, scientists either adopted agnostic attitudes, or kept their work separate from their religious beliefs and practices.

Up to the present day, many scientists hold the view that scientific investigation and religious thought apply to totally different and separate domains, and that their models and discoveries in the physical realm have little or no direct relationship to metaphysical, spiritual, or religious concepts. But by proceeding carefully, and applying the scientific method to consciousness in an objective manner, we may re-unite physical science with its proper metaphysical basis.

Transcendental Science: Ushering In A New Era

A new era is dawning. Scientists seeking to expand the envelope of theoretical and empirical methods in order to understand the nature of physical reality at the quantum and cosmological extremes of scale, have discovered radical evidence in their experimental results that legitimately raise important metaphysical questions. The time has come to close the circle, to recognize the importance of science's metaphysical basis, and to expand its area of application to include consciousness.

As we have demonstrated in the previous chapters, consciousness is a necessary element in the creation and propagation of the elementary particles that physics has discovered underlying physical reality. With the establishment and recognition of this fact, a number of fascinating questions and areas of research become legitimate concerns of science for the first time. For example, we can ask whether consciousness can exist apart from the structure and form of physical bodies. Is individual consciousness slowly evolving toward higher levels of cognition? Do subtle realities exist beyond the sensitivity and range of today's technological instrumentation, or are such things the imaginings of mystics and dreamers?

The Metaphysical Roots of Science

Contrary to what one might be led to think reading current scientific literature, many of the foremost scientists of the past century were not materialists. It is certainly no accident that some of those who produced the most important breakthroughs, some of the most revered scientists, were deep metaphysical thinkers, perhaps even mystics at heart. Albert Einstein is a prime example. Ken Wilber, in Quantum Questions, Shambala Publications, 1984, said of him: "Einstein's mysticism has been described as a cross between Spinoza and Pythagoras; there is a central order to the cosmos, an order that can be directly apprehended by the soul in mystical union."

But we can let Einstein speak for himself. In The New York Times Magazine, November 30, 1930, pages 1 - 4, he said:

> I maintain that the cosmic religious feeling is the strongest and noblest motive for scientific research. Only those who realize the immense efforts, and above all, the devotion without

which the pioneer work in theoretical science cannot be achieved are able to grasp the strength of the emotion out of which alone such work, remote as it is from the immediate realities of life, can issue. What a deep conviction of the rationality of the universe and what a yearning to understand, were it but a feeble reflection of the mind revealed in this world, Kepler and Newton must have had to enable them to spend years of solitary labor in disentangling the principles of celestial mechanics! Those whose acquaintance with scientific research is derived chiefly from its practical results easily develop a completely false notion of the mentality of men who, surrounded by a skeptical world, have shown the way to kindred spirits scattered wide through the world and the centuries. Only one who has devoted his life to similar ends can have a vivid realization of what has inspired these men and given them the strength to remain true to their purpose in spite of countless failures. It is cosmic religious feeling that gives a man such strength. A contemporary has said, not unjustly, that in this materialistic age of ours the serious scientific workers are the only profoundly religious people.

And in Mein Weltbilt, Querido Verlag, Amsterdam, 1934, he said:

> You will hardly find one among the profounder sort of scientific minds without a religious feeling of his own...the scientist is possessed by the sense of universal causation...His religious feeling takes the form of a rapturous amazement at the harmony of natural law, which reveals an intelligence of such superiority that, compared with it, all the systematic thinking and acting of human beings is an utterly insignificant reflection. This feeling is the guiding principle of his life and work, in so far as he succeeds in keeping himself from the shackles of selfish desire. It is beyond question closely akin to that which has possessed the religious geniuses of all ages.

Eight years younger than Einstein, Erwin Schrödinger was the author of the quantum wave equation, a more elegant expression of the same quantum reality described by Werner Heisenberg's probability matrices. His wave equation constitutes the basis of just about every modern textbook on quantum mechanics. While his most noted contributions are in the field of theoretical physics, his thinking has also influenced research in biology and other fields. The following quotes, revealing the surprising depth of his metaphysical and even mystical insight, are taken from essays published by the Cambridge University Press between 1947 and 1964.

> The same elements compose my mind and the world. this situation is the same for every mind and its world, in spite of the unfathomable abundance of "cross-references between them. The world is given to me only once, not one existing and one perceived. Subject and object are only one. The barrier between them cannot be said to have broken down as a result of recent experience in the physical sciences, [specifically the effects of observation in quantum mechanics experiments...] for this barrier does not exist.

Concerning the relationship of conscious beings to the whole of reality, he said:

> Suppose you are sitting on a bench by a path in high mountain country. There are grassy slopes all around, with rocks thrusting through them; on the opposite slope of the valley there is a stretch of scree with a low growth of alder bushes. Woods climb steeply on both sides of the valley, up to the line of treeless pasture; facing you, soaring up from the depths of the valley is the mighty glacier-tipped peak, its smooth snow fields and hard-edged rock faces touched at this moment with soft rose colour by the last rays of the departing sun, all marvelously sharp against the clear, pale, transparent blue of the sky.

According to our usual way of looking at it, everything that you are seeing has, apart from small changes, been there for thousands of years before you. After a while - not long - you will no longer exist, and the woods and rocks and sky will continue, unchanged, for thousands of years after you.

What is it that has called you so suddenly out of nothingness to enjoy for a brief while a spectacle which remains quite indifferent to you? The conditions for your existence are almost as old as the rocks. For thousands of years men have striven and suffered and begotten and women have brought forth in pain. A hundred years ago, perhaps, another man sat on this spot; like you, he gazed with awe and yearning in his heart at the dying light on the glaciers. Like you, he was begotten of man and born of woman. He felt pain and brief joy as you do. Was he someone else? Was it not you yourself? What is this self of yours? What was the necessary condition for making the thing conceived this time into *you*, just *you*, and not someone else? ...What justifies you in obstinately discovering this difference - the difference between you and someone else - when objectively, what is there is *the same*?

Looking and thinking in that manner, you may suddenly come to see, in a flash, the profound

> rightness of the basic conviction in Vedanta [a basically Hindu world view, taught in the West by Swami Vivekananda, disciple of Ramakrishna]: it is not possible that this unity of knowledge, feeling, and choice which you call your own should have sprung into being from nothingness at a given moment not so long ago; rather this knowledge,feeling, and choice are essentially eternal and unchangeable and numerically *one* in all men, nay in all sensitive beings.

And what about Max Planck, discoverer of the quantum nature of matter and energy? What did he think about the relationship between science and religion, between the exact knowledge of science and religious feeling? In an essay entitled Where is Science Going?, Norton, New York, 1932, he had this to say:

> There can never be any real opposition between religion and science. Every serious and reflective person realizes, I think, that the religious element in his nature must be recognized and cultivated if all the powers of the human soul are to act together in perfect balance and harmony. And indeed, it was not by any accident that the greatest thinkers of all ages were also deeply religious souls, even though they made no public show of their religious feeling. It is from the cooperation of

> the understanding with the will that the finest fruit of philosophy has arisen, namely the ethical fruit. Science enhances the moral values of life because it furthers a love of truth and reverence - love of truth displaying itself in the constant endeavor to arrive at a more exact knowledge of the world of mind and matter around us, and reverence, because every advance in knowledge brings us face to face with the mystery of our own being.

The Cosmic Religious Feeling

Science and religion, properly understood, should not be in conflict; they spring from the same noble impulse, the desire: to know. This impulse is described by Einstein and others as the "cosmic religious feeling". Religion brings meaning to science. Science without spirit is dead; and yet, many scientists profess to believe that science and religion exist in totally separate domains. They hold that the use of scientific objectivity and logical proof is generally limited to the investigation of what is currently recognized as physical reality. Even those who feel that objective reasoning may be applied to the realm of consciousness and spirit, usually regard this realm as more or less separate from the physical universe. It is here that we must take exception and part company with them. If science is to advance beyond the limitations of gross materialism, we must shed this mental fixation and move beyond it. While it was necessary and appropriate to the last century and perhaps the first half of this one, scientific materialism is dead. It will become a millstone around the necks of those who cling to it, preventing the birth and growth of a much-needed, broader-based transcendental science.

Science had to divorce itself from religion and superstition, and in order to do so, it had to focus on the more provable, material aspects of reality. But we have come a long way from the days when science was called natural philosophy, to today's genetic experimentation, virtual reality, high-tech warfare, and relativistic-quantum physics. But what does science have to say about the nature of reality and our relation to it? Is consciousness an accidental byproduct of the evolution of the physical universe, or is the entire cosmos, from quarks to stars, created and sustained by the activities of consciousness? We now know the answer to this question. We no longer have to base our answer on appearances and theoretical reconstructions of the evolution of the physical universe. We know that many aspects of physical reality, including quantum particles and waves, time, and space are selected out of a much broader spectrum of reality by the action of conscious observation. The form and structure of physical reality are results of the functioning of consciousness.

Beyond Materialism

With proof that the attitude of scientific materialism is no longer viable as a basis for scientific research, we can explore the hypothesis that consciousness is a real force in the universe. We need to form a new, broader theoretical basis for understanding reality that includes the functioning of consciousness in relation to the creation and propagation of logical structure and order in the universe, and we need a new mathematical language designed to describe it.

Objectivity must now be applied to more than just the traditional concepts of physical reality. With the Aspect experiment, we now have empirical evidence that consciousness exists as a real and objective part of reality, actively involved in the creation and sustenance of the form and structure of the universe. We know now that physical reality would not exist in the forms that we behold without the functioning of both primary and individual forms of consciousness.

Scientific Dogma

Science and religion are soul mates. Both arise from human curiosity, the desire to know the truth. Both are grounded in metaphysics. Both seek to enlighten us and improve our general well being. In many ways, the current scientific establishment is like a subtle form of priesthood, ostensibly existing to guide us to truth and enlightenment, but more often than we would like to believe, functioning to obscure the truth and keep us dependent upon the specialized knowledge and expertise of an elite few. It is no longer a religious priesthood, but a scientific illuminati who claim to know the only path to truth. They believe, just as their sectarian predecessors did, that they, and <u>only</u> they, hold the keys to truth and knowledge. They defend and guard their belief systems just as zealously as any priesthood ever defended religious dogma. But the answer to the question "What is truth?" should not depend upon religious or scientific doctrine. We must not refrain from asking certain questions just because they are not considered by the current authorities to be scientifically or politically correct.

Because of the a priori assumption of the absolute separation of consciousness and matter, the current politically and scientifically correct view is that consciousness is an emergent feature of the evolution of the physical universe. Adherence to this position, however, is no longer viable with the advent of quantum mechanics and the results of the Aspect experiment. With empirical evidence and proof that particles and waves, the building blocks of everything in the physical universe, cannot exist without the involvement of consciousness the a priori assumption of absolute separation must be discarded. The question of the role of consciousness in the creation of man and the universe is once again open for scientific investigation.

New Questions, New Answers

Does conscious observation create particles and waves? If so, is there an objective reality independent of consciousness? If there is, it is clearly unlike the universe we perceive and have come to regard as objective. The universe that we perceive depends upon the existence of particles and waves. And now, thanks to Bell's theorem and the Aspect experiment we know that what we perceive as physical reality is either nonlocal, nonobjective, or both. Time and space, and to some extent, even matter and energy, are revealed as artifacts of consciousness.

Quarks and Beyond

Since the time of Rutherford, we have continued to search for the most basic building blocks of physical reality. But the ultimate particle kept slipping through our fingers for a long time. Solid matter turned out to be mostly empty space. Atoms gave way to a plethora of sub-atomic particles, made up of yet smaller particles called hadrons, made up of quarks, the most substantial of which are elusive and ephemeral, to say the least. Recently some physicists have proposed that quarks may actually be tiny, vibrating string-like structures. Traces of the final of the 12 quarks, the top quark, were only finally captured as recently as the spring of 1995. And now we know that something else exists beyond the smallest quark, and whatever their structure, particles like photons and electrons do not exist until they impact on a receptor.

This "something else" beyond the quantum is the beginning of the trail leading to an understanding of consciousness; but, trying to grasp the essence of consciousness may prove to be even more difficult than chasing the ultimate particle. The dialectic you have been given with which to think about yourself and the reality you experience is cast in terms of self and other, subject and object. You have a sense of self, but when you attempt to define it in terms of known objects, you find that, no matter what you describe, it is always something other than self. You are always that which perceives, never the object of perception. As soon as you think of some organ, perhaps eye or brain, as the seat of perception, that organ becomes an object, and you must ask: what is it that perceives this object? Trying to grasp the root of your own consciousness, you find that it dissolves...and the abyss opens to receive you.

We have now determined that something exists within each one of us that is not composed of matter, atoms, quarks, strings, or anything like them. Empirical evidence and logic attest to the fact that it exists, and the only way we know reality, the world, and the universe is through it. We've described it as a nexus of mind and matter, a receptorium. Is this a gateway to the direct perception of a broader spectrum of reality than our physical senses can access? Is it possible that the receptorium can be expanded in some way to intercept more of the reality spectrum? Is there some way we can access it directly? Religious leaders like Jesus, Buddha, Patanjali, and Mohammed, and many mystics, both ancient and modern, have attested to the fact that we can.

Mystics and Science

Those who attempt to understand reality through inner exploration are generally branded as mystics by today's science. What do we mean by the term mystic? Was Erwin Schrödinger a mystic? Did

Einstein have mystical leanings? Could real mystics simply be people who have located the receptorium and explored it --and beyond? Unfortunately, the term mystic is so misunderstood and misused by modern writers as to be nearly useless. Even Webster's definitions reveal the confusion that has been fostered by the pervasive attitude of scientific materialism:

> Mystic, *adverb*: Hidden from or obscure to human knowledge or comprehension; *noun*: A believer in mysticism, one who professes to have mystical religious experiences: one initiated or admitted to occult rites.
>
> Mysticism: The theory or belief that man can intuitively know God or religious truth, through the inward perception of the mind, a more immediate and direct method than that of ordinary understanding or sense perception; any seeking to solve the mysteries of existence by internal illumination or special revelation; a dreamy contemplation or speculation on ideas that have no foundation in human experience.

These dictionary definitions do reveal a definite materialistic bias, however, they are dual in nature. They allow for the belief in God and religious truth, but suggest that mystics' claim access to knowledge that may be fantasy or mere speculation, far removed from real human experience or comprehension. Even science writers who refer to the mysticism of some of the most important scientists often imply that any metaphysical basis for experiment, thought or theory is mystical and that reference to anything not included in the current physical description of the universe has no foundation in objective reality and thus has no place in intellectual discussion. The world's great mystics: Jesus, Buddha, Mohammed, Kabir, Ramakrishna, Meister Eckhart, and many others, however, have attested to the reality of human mystical experience, and insist that transcendental understanding is available to everyone.

Objectivity No Longer Limited to the Material World

Objectivity is a legitimate aim of science, but the use of this term also suffers from the same bias as mystic and mysticism. Webster says that objectivity is "...intentness on objects independent of the mind; reality outside the mind." How did we come to believe that there are no objects, objectivity, or reality within the mind, or within consciousness? It appears that the over-simplification of scientific materialism has impacted the effectiveness of our language. This narrow definition of objectivity excludes everything beyond that which we can sense, weigh, or measure with our current technology, and thus restricts what we are supposed to think of as real. It is time for scientists to be objective about objectivity. A more comprehensive definition of objective reality must include the whole spectrum of phenomena: from the gross forms of matter and energy to more and more subtle forms, including those of consciousness.

There are objective internal phenomena as well as objective external phenomena. They exist at opposite ends of the spectrum of reality. Internal objectivity is at least as necessary and important to the advancement of knowledge and understanding as external objectivity. It is therefore important to recognize that there may be true mystics who are not dreamers living in fantasy worlds. Their function may be to bring the realities of consciousness more fully into the physical world. They may be aware of aspects of reality far beyond the common experience. They may even be in tune with the natural flow of form and structure from pure consciousness into physical manifestation.

Objectivity, Mystics, and Miracles

There are numerous instances of credible witnesses who have reported experiencing extraordinary, or even miraculous events, under certain circumstances, in certain locations, or in the presence of certain individuals. Science has generally ignored such reports, dismissing them as anecdotal and subjective. However, they have persisted from before the time of Jesus right down to the present. Miracles were reported in connection with Padre Pio in Italy and many people reported seeing him levitate while in prayer or meditation. Paramahansa Yogananda, founder of Self-Realization Fellowship in Los Angeles, is reported to have read the thoughts of people, brought about healings and other miracles. Many report that Satya Sai Baba materializes objects from thin air in Southern India. If even a few of such reports are legitimate, it is possible that they were effects of the operation of some sort of higher consciousness. With the Aspect results and the proof of the reality of consciousness beyond the quantum, we now have at least a scientific basis for considering and investigating such phenomena.

Through a serious study of genuine mystics of the past, we can see that the true mystics of all ages have taught transcendence of illusion, pain, and suffering by contact and merger with a higher form of consciousness. They urge us to try to grasp the root of being, the significance of consciousness and life, the real relationship of consciousness to matter. The true mystic does not pursue a "dreamy contemplation or speculation on ideas that have no foundation in human experience." On the contrary, the true mystic pursues the very essence of the experience of reality. When Einstein said: "I want to know God's thoughts, the rest are details." He wasn't indulging in dreamy contemplation. He was serious.

The Non-Quantum Receptorium and Mystical Experience

If matter, energy, time, and space, as we know them, are dependent upon consciousness, then the key to understanding the nature of reality lies in understanding the nature of consciousness. Each one of us experiences consciousness; but do we all experience the same thing? If Erwin Schrödinger was right, and the "I" that each one of us experiences is the same, "numerically one", then the individual differences may be differences of degree only, not differences of kind. Science is now at last at the point where it can move beyond the ultimate particle toward an understanding of the receptorium, the gateway into consciousness.

There is evidence that I find to be at least as convincing as the evidence for the existence of quarks, that real mystics, with deep understanding of reality do exist. They appear to be in contact with reality on a deeper level that allows them to accomplish things that seem impossible to the average person. Just about 100 years ago, a Canadian doctor named Richard M. Bucke coined the term "Cosmic Consciousness". He published a book by that title and put forth the theory that consciousness expresses itself in the physical realm in progressively sophisticated forms, ranging from the bare awareness, present in simple organisms, to self-awareness in human beings, to a transcendental awareness, which he called Cosmic Consciousness. In the opening section of the book, which he entitled "First Words", Bucke defined Cosmic Consciousness in this manner:

> Cosmic Consciousness is a third form which is as far above Self Consciousness as is that above Simple Consciousness. With this form, of course, both Simple and Self Consciousness persist (as Simple Consciousness persists when Self Consciousness is acquired), but added to them is a new faculty so often named and to be

> named in this volume. The prime characteristic of Cosmic Consciousness is, as its name implies, a consciousness of the life and order of the universe. ... Along with the consciousness of the cosmos there occurs an intellectual enlightenment or illumination which alone would place the individual on a new plane of existence - would make him almost a member of a new species. To this is added a state of moral exaltation, an indescribable feeling of elevation, elation, and joyousness, and a quickening of the moral sense, which is fully as striking and more important both to the individual and the race than is the enhanced intellectual power. With these come, what may be called a sense of immortality, a consciousness of eternal life, not a conviction that he shall have this, but the consciousness that he has it already.

Bucke presents what was known of historical figures who, in his opinion, were instances of Cosmic Consciousness in the flesh. The list includes Jesus Christ, Gautama Buddha, Paul, Mohammed, Walt Whitman, and many others. In the final section, "Last Words", he concludes:

> ...a Cosmic Conscious race will not be the race which exists today, any more than the present race of men is the same race which existed prior to the evolution of self consciousness.

> The simple truth is, that there has lived on the earth "appearing at intervals" for thousands of years, among ordinary men, the first faint beginnings of another race; walking the earth and breathing the same air with us ... This new race is in the act of being born from us, and in the near future it will occupy and possess the earth.

Primary Consciousness and Cosmic Consciousness

Whether or not Bucke has made the case for Cosmic Consciousness, there is evidence that a more sophisticated form of consciousness may exist. The results of the Aspect experiment imply that a form of consciousness that is nonlocal had to exist prior to the existence of matter and individual consciousness. This consciousness must contain all of the intricate structure apparent in the physical universe, and perhaps more. If time and space are artifacts of the functioning of consciousness, then whatever intelligence, order, and meaningful structure exist now, or can ever exist, may already exist in this primary form of consciousness. Spiritually advanced or "enlightened" beings, such as Jesus may simply be aware of and rooted in this primary form of consciousness.

The following excerpt from "Autobiography of a Yogi" by Paramahansa Yogananda is offered as a description of the state of consciousness Bucke referred to as cosmic consciousness. Yogananda was not writing as a poet, in imaginative or speculative terms, but as a highly trained and evolved mystic, reporting an actual experience. He tells us that this was his first experience of Cosmic Consciousness, under the guidance of his spiritual teacher:

> My body became immovably rooted; breath was drawn out of my lungs as if by some huge magnet. Soul and mind instantly lost their physical bondage and streamed out like a fluid piercing light from my every pore. The flesh was as though dead; yet in my intense awareness, I knew that I had never before been fully alive. My sense of identity was no longer narrowly confined to a body but embraced the circumambient atoms. People on distant streets seemed to be moving gently over my own remote periphery. The roots of plants and trees appeared through a dim transparency of the soil; I discerned the inward flow of their sap.
>
> The whole vicinity lay bare before me. My ordinary frontal vision was now changed to a vast spherical sight, simultaneously all-perceptive. Through the back of my head I saw men strolling far down Rai Ghat Lane, and noticed also a white cow that was leisurely approaching. When she reached the open ashram gate, I observed her as though with my two physical eyes. After she had passed behind the brick wall of the courtyard, I saw her clearly still. All objects within my panoramic

gaze trembled and vibrated like quick motion pictures. My body, Master's, the pillared courtyard, the furniture and floor, the trees and sunshine, occasionally became violently agitated, until all melted into a luminescent sea; even as sugar crystals, thrown into a glass of water, dissolve after being shaken. The unifying light alternated with materializations of form, the metamorphoses revealing the law of cause and effect in creation.

An oceanic joy broke upon calm endless shores of my soul. The spirit of God, I realized, is exhaustless bliss; His body is countless tissues of light. A swelling glory within me began to envelop towns, continents, the earth, solar and stellar systems, tenuous nebulae, and floating universes. The entire cosmos, gently luminous, like a city seen afar at night, glimmered within the infinitude of my being.The dazzling light beyond the sharply etched global outlines faded slightly at the farthest edges; there I saw a mellow radiance, ever undiminished. It was indescribably subtle; the planetary pictures were formed of a grosser light.

The divine dispersion of rays poured from an eternal source, blazing into galaxies, transfigured with ineffable auras. Again and again, I saw the creative beams condense into constellations, then resolve into sheets of

> transparent flame. By rhythmic reversion, sextillion worlds passed into diaphanous luster, then fire became firmament.
>
> I cognized the center of the empyrean as a point of intuitive perception in my heart. Irradiating splendor issued from my nucleus to every part of the universal structure. Blissful *amrita*, nectar of immortality pulsated through me with a quicksilverlike fluidity. The creative voice of God I heard resounding as *aum*, the vibration of the Cosmic Motor.

The history of every major culture on this planet has recorded evidence of enlightened beings, called prophets, seers, or saints, according to the language and understanding of their time and place. While ponderous religious institutions grew up in the footsteps of many of them, their real purpose has been to inspire and uplift us, to encourage us to aspire to higher standards and greater things than those into which we find ourselves cast at birth.

The Negative Impact of Materialistic Science

Ironically, the wholesale acceptance of materialism as the basis of all science has gradually brought about defensive, unscientific attitudes in many modern scientists, very similar to those exhibited by many religionists at the beginning of the ascendance of science. Raising transcendental concepts in relation to scientific inquiry often elicits attitudes of anger in individuals who apparently feel the validity of the metaphysical bases of their own theories and world views threatened. They act as if they must defend their territory. But real scientists should have no need to defend their territory, for real science has no boundaries.

Dogged adherence to materialism also has a stultifying effect on the mind; especially on intuition and creativity. Even though they may be intellectually gifted, scientists who exhaust the scope of the accepted materialistic models adopt limited, fatalistic views concerning the ends of science and humanity in general. Having explored all the options of their closed materialistic model, they think they have discovered all there is to know about the nature of reality. The French mathematician La Place, for example, saw no need for the idea of God. In the early 1700's he declared that it would only take a few years for scientists to determine the initial conditions of the universe, after which its complete history and fate could be calculated using Newtonian mechanics. More recently, Stephen Hawking, Lucasian Professor of Mathematics at Cambridge, speculated that a "theory of everything", incorporating relativity and quantum physics may be in hand by the end of the century. At that time, he suggests, we will have the means to know all there is to know about the universe, from the big bang to the big crunch at the end of time. These brilliant men have made the simple error of confusing their models with reality. In forming their basic assumptions, they have either totally overlooked, or at least, grossly underestimated the impact and importance of the functioning of consciousness in the universe.

The prime example of the extreme pessimism of scientific materialism is Bertrand Russell, one of the foremost proponents of materialism, who said in A Free Man's Worship that all the works of man, however wonderful, are "destined to extinction in the vast death of the solar system". And in 1977, Nobel Prize-winning physicist Steven Weinberg said that "the more the universe seems comprehensible, the more it seems pointless". These materialistic scientists are like travelers using a faulty map. Having lost their way, they blame the landscape for their confusion. Once again, they are mistaking their materialistic model for reality - a reality we now know contains far more than their models allow.

By far the most serious impact of scientific materialism that has filtered down to the man on the street is the loss of belief in the potential divinity of man. The transcendental nature of the beings possessed with Cosmic Consciousness, discussed by Doctor Bucke, is a meaningless myth to the average citizen who has bought into the materialistic belief system promoted by the current paradigm. The explosion of youth gangs and their violent behavior is a clear symptom of this loss of belief in the wonder of life and the miracle of creation. The value of life is lost on those who cannot see beyond the goals of satisfying physical desire. This trend can only be reversed by changing the modern world view projected by science and technology and perpetuated by Hollywood - Madison Avenue commercialism.

A science based on materialism is necessarily finite, limited, and closed to the real potentials of human consciousness. We know now that the universe is not limited to the sphere that current science defines as physical. Not only does reality include an active function called consciousness, but this function arises from a subtle form underlying the substance that makes up the elements of the limited sphere now known as the physical universe. Furthermore, we know, thanks to Einstein, Bohr, Bell, and Aspect, that the functioning of this subtle form is a pre-requisite to the existence of the physical universe.

The belief that reality is meaningless is very dangerous because it suggests that ethical considerations are pointless. Consciousness is assumed by materialistic science to be an abstract epiphenomenon resulting from a chance combination of physical factors. But the results of Bell's theorem and the Aspect experiment imply that this is not true. Seen in the light of current events, the question: "What is the nature of reality?" becomes ever more important. We can no longer afford to assume that an amoral approach to science is sufficient and appropriate.

We think that the crime and violence sweeping our land today are caused by racism, prejudice, insensitivity and hatred. These, however, are symptoms of a deeper cause: the loss of meaning. Many people today, when asked: "What is the true nature of reality?" and "What is the meaning of life?"

have no answers at all. Many feel that we are adrift on the ocean of life without purpose or direction. There may well be a correlation between the increasing tide of senseless violence and the brutal message behind most television drama and the evening news, bristling with negativity and violence, that there are no absolutes, no meaning to life. How much of this is a reflection of the underlying materialistic assumptions of modern science and technology? We are educating our youth to be computer literate and politically correct, while stifling their inner growth by teaching them that life has no meaning beyond money and sex. Because of the tacit assumption of scientific materialism, we find leading scientists mistaking their materialistic models for reality and declaring that reality is devoid of meaning or purpose. The technological ability to dominate and destroy has been mistaken for progress and the possession of factual information has been confused with knowledge and understanding.

Reversing the Trend

Science has become such an important influence in modern life that it has replaced religion as the keeper of truth. As long as science recognizes no substance or reality beyond the quanta of the physical universe, and no purpose or meaning behind them, this emptiness will be transmitted through commercial technology to a despairing populace. Fortunately, this no longer has to be the case. With the recognition of the empirical evidence of the Aspect experiment and the logical necessity of a primary form of consciousness prior to the manifestation of any physical object, science can regain the meaning lost with the adoption of materialism as a metaphysical basis for investigation. The trend toward self-destructive behavior, a natural product of the loss of meaning, can yet be reversed.

In the chapters to come, we will develop the assumptions, logic, and structure of a new paradigm of transcendental physics embracing consciousness and nonlocal reality. Such a vital new theory is urgently needed to bring us back to the tradition of Newton and Einstein so that we can move on from the current sterile, materialistic technology to a broader, consciousness-based science, which will be capable of integrating all of our efforts to search for truth into one single, effective, meaningful scientific paradigm.

CHAPTER 5.
LIGHT AND THE PORTALS OF CONSCIOUSNESS

> And God said let there be light: and there was light. And God saw the light, that *it was* good: and God divided the light from the darkness.
>
> - Genesis, Chapter 1, verses 3 & 4

Light is the medium of observation. Without the physical phenomenon of light, we would have little knowledge of any external world. Beyond that, the perception of light is so central to our understanding that it is linked linguistically with truth and revelation. Automatically, our consciousness constructs mental images of scenes analogous to those we perceive to exist in the physical world in terms of contrasting shades of light. Even in the dark of night, with eyes closed in sleep, we may see scenes bathed in light and color. And many people report experiencing dreams or visions of light that appear to be more vivid and more alive than the lights of the external world.

Light conveys truth, but it can also convey illusion. Most religions hold the view that the world we experience is, at least to some extent, illusory. Can it be that science, the bastion of realism, is in the process of proving them right? We seem to see solid forms of mountains and rivers rising and falling, plants and animals, living and dying under the forces and laws of nature. But the physical forms are not solid; far from it, they are mostly empty space. We seem to see sunlight and shadow, the sun and moon, massive bodies of matter and energy. But now we find that we don't actually see matter or energy, we only see the effects that elementary quanta create on receiving structures. And time and space have no meaning without conscious perception and memory, and thus, for all we know, may not even exist without the functioning of consciousness. The common sense and logic of science, pursuing the truth of atomic structure, have brought us to the doors of perception where common sense seems no longer to apply. Common sense and common wisdom, it seems, are in need of revision.

We have likened the development of science to the flow of a river. Contrary to common wisdom, water does not always run down hill. Along the course of a river there are areas of reverse flow, circular eddies, and at times of flood, hydraulic surge or jump can occur, when water will actually flow up hill. So it is with the stream of scientific thought. We like to think of science as always progressing, always correct, continuously adding to the pool of existing knowledge. But so far, in the history of science, every paradigm devised by man has eventually proved to be at least partially incorrect, and without exception, they have been incomplete.

There are many examples of ebb and flow in the succession of scientific theories, but none more central to physical science and to our discussion, than the theory of the nature and propagation of electromagnetic radiation, the family of vibratory energy that includes light, the everyday medium of all our perceptions of physical reality.

In 1690, the Dutch physicist Christian Huygens advanced the theory that light is propagated in spherical waves through a subtle, all-pervasive universal substance called aether. Sir Isaac Newton took issue with this theory and in 1704 published his own theory holding that light is made up of innumerable very small particles, which he called corpuscles. It wasn't until almost a century later that the tide swung the other way again, when Thomas Young demonstrated with his double slit experiment in 1801, that light behaves in ways that can only be explained by a theory of oscillating waves. The wave theory of light became well ensconced after 1867, when James Clerke Maxwell introduced the electromagnetic field theory. With the success of Maxwell's wave equation, science considered the matter settled. Light definitely appeared to be a wave phenomenon.

In 1905, Einstein, inspired by Planck's discovery of the quantum nature of black-body radiation, published a paper explaining the photoelectric effect (an electric current induced by shining light on metal) by assuming that light behaves as discreet quanta interacting with the atoms of certain metallic surfaces, causing a flow of electrons, each photon of light energy producing an electron. It wasn't until 1923 that this theory was confirmed experimentally. With this confirmation, physicists accepted something that would have been thought impossible or even ridiculous in Newton's day: light can behave as either wave or particle.

In his doctoral thesis, written in 1924, Louis de Broglie suggested that electrons might share this characteristic of light and behave as either particle or wave. Scientists, with the possible exception of Einstein, didn't take his conjecture very seriously at the time. But, with the development of quantum mechanics, specifically Schrödinger's wave equation, it became apparent that all quantum particles have associated wave lengths. It turns out that particles and waves are not mutually exclusive, but actually complementary aspects of all elementary particles, including photons.

But light is an even more complex phenomenon than that. As we saw in the previous chapter, the Copenhagen interpretation tells us that photons do not exist until an observation or measurement is made. We know that the observed effects of light can be attributed to either waves or particles, and we have seen in the two-slit experiment, that there is at least one way that an observer can consciously choose to cause a beam of light to manifest as one or the other. Since light is the medium of observation, our notions of time and space are intimately related to the propagation and observation of light. With what we know now, it appears that our theories, devised to explain various properties and aspects of light are still only superficial descriptions of the tip of the iceberg.

There is one feature of light that can actually help us to understand the Copenhagen interpretation of quantum phenomena. After all, light *is* a quantum phenomenon, at least when it manifests as photons. Think of how a full moon looks on a cloudless night. Its rocky surface is bathed in bright light, its edges etched starkly white against the blackness of space. But the sun does not focus its rays so that they strike only on the surfaces of the moon, earth, and the other planets and satellites. If light were a localized physical phenomenon, the space around the moon would be filled with the same density of light that strikes the moon, but we don't see it there.

With the Copenhagen interpretation confirmed by the results of the Aspect experiment, we can now say without fear of contradiction, that no one has ever seen a photon, since photons don't exist as particles or waves in flight. We can only see the effects, after light has registered on a receptor. So we see that light, something every-day familiar, behaves just as Bohr said all quantum particles do: it doesn't exist until it registers on a receptor in a such a way that its effects can be observed or measured.

Before Bell's theorem and the Aspect experiment, we had to believe that photons existed between the sun and the moon, and between any source and receiving object, we just accepted the fact that we couldn't see them until they struck an object and were reflected to our eyes. But, even though we believed that they existed in flight, there was no way we could actually prove that they did, because the only way we could detect them was by placing an object, a receptor, in their path. We know now that they don't exist until they strike an object. The EPR paradox destroys the quantum theory if this is not so, and Bell's theorem and the Aspect experiment proved that it is so. So no one has ever seen light! You might argue that you have seen beams of light streaming down from a hole in the clouds, or through a window and across a room. But you have not seen photons. You have not seen light. You have only seen the effects of light quanta on particles of moisture or dust in the air.

Thanks to quantum mechanics, Bell's theorem, and Aspect's empirical results, we now have a totally new and different understanding of the nature of reality. And, as we have seen, this new understanding includes light, which means that it affects all our observations. This new understanding has the potential of transforming all our perceptions, even our perceptions of consciousness and of our selves.

As we investigate the question of how observation causes light to manifest various aspects of its complex nature, it will be instructive to look at light from the perspective of Einstein's relativity. For the reader without training in relativistic physics, who would like to pursue it, Einstein's little book Relativity, the Special and the General Theory, A Clear Explanation That Anyone Can Understand, Crown Publishers, New York, 1961 is recommended as a basis for the following discussion.

For the purposes of this visualization, we will first return to the thinking that was prevalent before the Aspect experiment confirmed the Copenhagen interpretation and accept the following basic Einsteinian assumptions: 1.) Light consists of quanta of energy that traverse every point of straight-line locality between their source and their eventual point of impact and absorption; 2.) There is no preferred reference frame; and 3.) these quanta always travel at **c**, the speed of light, relative to any observer regardless of relative motion, and nothing can travel faster than **c**.

Now consider two conscious observers moving toward each other at a very high rate of speed, as depicted in Figure 5. They are moving on near-collision, straight-line parallel courses at uniform velocities. They pass a light source at the same time they pass each other. A straight line drawn from one observer to the other will pass through the light source and will be perpendicular to the observers' lines of travel when the observers pass. (See Figure 5.) Further, assume that the light source is emitting one photon at a time.

According to the theory of relativity, either of the observers, looking into the other's reference frame, will see the other's space contracted along the line of motion. Each unit of length will be contracted by a factor that may be calculated using the Lorentz contraction equation. The Lorentz contraction equation is easily derived using nothing more than high-school algebra, geometry, and physics. A simple derivation of this type is peresented in Appendix E-I.

If observer **A** calculates the speed of a photon in the contracted space of the other reference frame, due

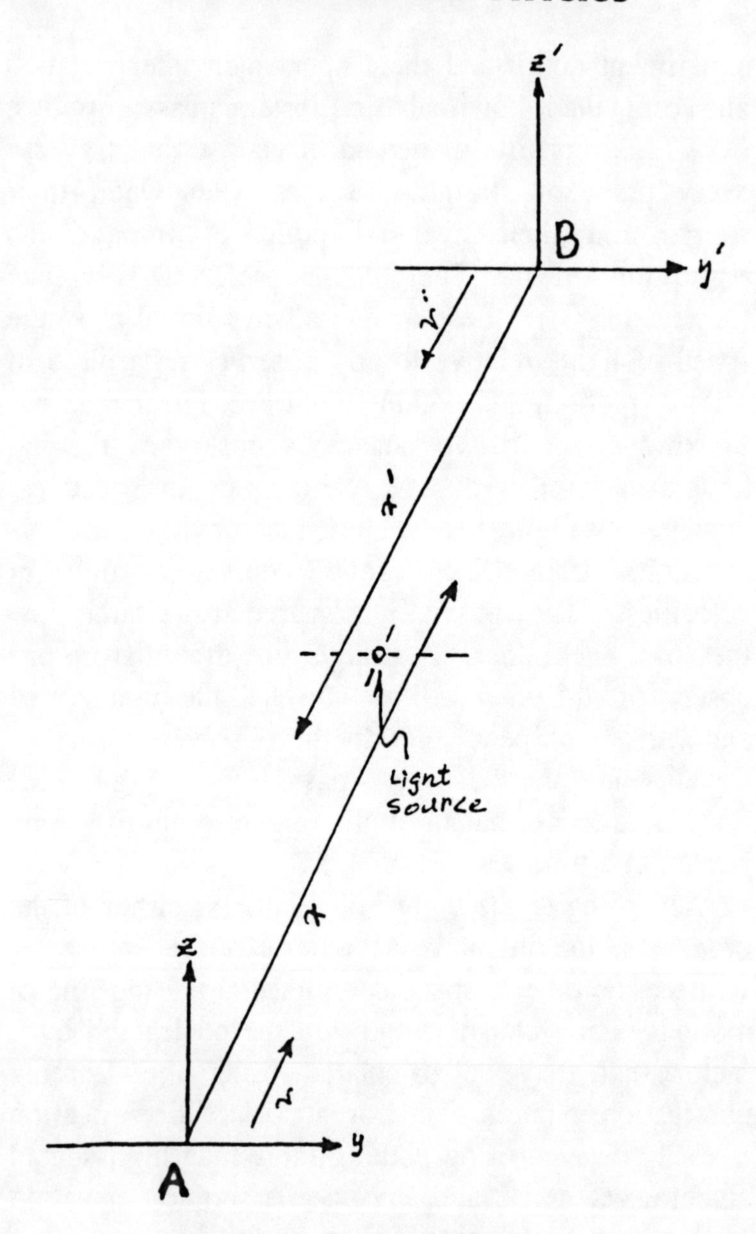

Figure 5. Relative Motion

to the requirement of assumption 3, time in the contracted reference frame must be passing more slowly than in the observer's frame in order for the photon to be travelling at **c**, the one and only speed of light. The amount that time must slow down in the contracted reference frame can be calculated using the Lorentz time dilation equation. If we think in terms of a four-dimensional time-space continuum, transformation from a point in the space-time of observer **A** to the corresponding space-time point in the other reference frame may be accomplished using the Lorentz transformation equations. However, if we transform from that point in the other reference frame back to reference frame **A**, we find that it is not the same point we originally transformed. For a thorough treatment of this point, with calculations of the transformations of space-time points from each observer's reference frame to the other, see Appendix E-II, page 331.

While a given photon can only be intercepted by one of the observers in Einstein's universe of local reality, if it is a simple physical object, then at any point in time and space it must be available for interception by either. But it is impossible for both observers to see the same physical object if its location cannot be transformed both ways unambiguously. What could the nature of light be that would allow it to behave this way? Can light start as a particle at the source, expand to exist as a wave front while travelling through space, and condense into a particle again upon impact? This might explain the interference patterns of the two-slit experiment, but it can't explain the results of the delayed-choice experiment, and it violates two of the assumptions of relativity: 1.) There is no photon traversing every point along the line from source to receptor, violating the assumption of locality; and 2.) The instantaneous expansion from the point of origin to a spherical wave form, and the subsequent instantaneous collapse of the wave from some extended area in space to a tiny pin point on the receiving surface violates the requirement that light energy always moves at the speed **c**.

We have to conclude that there is no place in the theory of relativity for a description of the photon. It cannot have a lateral dimension, since it is travelling at **c** relative to any observer; yet we know that it is a physical entity. It has measurable effects in the physical world. If it is a physical particle, or a physical wave, or a combination of the two, under the assumptions of relativity, it must behave to suit the space-time requirements of every possible observer: it must show up travelling at **c**, regardless of relative motion. Therefore, although dimensionless, it must carry with it any number of potential manifestations, to satisfy any moving observer it might chance upon. Does this sound familiar?

The Copenhagen interpretation again comes to the rescue: The photon does not exist as a tiny missile moving through every point of local space along its path -- in fact, it doesn't have a path. It exists in local, observational space-time only when its impact produces effects that are measurable and observable.

The photon is only a potential until a conscious observer evokes a manifestation of that potential by the act of the drawing of a distinction, i.e., making an observation. This fact makes light, if not the most important, clearly a very important link between individual consciousness and objective physical reality.

What the Copenhagen interpretation and the Aspect experiment tell us, when combined with relativity and the two-slit experiment, is that reality reveals to the conscious observer only the forms that are made distinct from the spectrum of possible forms by the methods of observation available to the observer. These forms appear to reflect a logical structure already existing in the observer's mind. While this implies that, as conscious observers, we participate in the unfolding of reality, it does not mean that we may create reality any way we please. We are constrained by a universal logical structure that seems to be more sophisticated and complex than we can imagine. But one thing is certain, it is impossible for a reality limited to the known forms of matter, energy, time and space to respond in this way. None of the known laws of physics explain this behavior, any more than they explain the existence of consciousness.

Clearly, reality is much more complex than our perceptions have led us to believe up to now. To behave the way it does, reality must be multi-dimensional, containing all the "probable" states at once, and what our senses, even extended by means of the apparatus we have constructed, allow us to perceive is only a small fraction of reality.

One of the most basic functions of consciousness is the drawing of distinctions. We might argue that the distinction of self from other is the first distinction a conscious entity makes. If this is so, the distinction of light from darkness must surely be the second. The structure we perceive in physical reality enters our consciousness primarily through the medium of light. If we wish to know and understand the nature of reality, it is imperative that we understand how and where this structure originates.

Michael Talbot said that we must be "prepared to consider radically new views of reality." Roger Penrose stated that "our very notions of actuality have become profoundly disturbed." And John Wheeler declared that "There is a strange sense in which this is a participatory universe." Wheeler is not the first physicist to think of the implications of quantum theory as strange. Werner Heisenberg in Physics and Philosophy, Harper & Row, New York, 1962, said: "I repeated to myself again and again the question: can nature possibly be as absurd as it seemed to us in these atomic experiments?" During the EPR debates Bohr said that a photon does not have a pathway; that the very concept of path is ambiguous when applied to quanta and that those who are not shocked by quantum theory have not understood it. After a lengthy debate with Bohr, Schrödinger, reluctant to accept the strange new world of pathless particles, is reported to have said that he was sorry that he ever started to work on atomic theory.

Why do even the scientists who developed quantum mechanics find its implications to be so strange? Perhaps it is because we know, at least subconsciously, that quantum observations contradict the time-honored assumption of the absolute separation of matter and consciousness. It is clear that such concepts run counter to the materialistic bias of modern science.

Science and technology have been pretty successful defining the gross characteristics of reality at the level immediately available to the senses; but at the extreme ends of the scale, at the quantum level and at the cosmological level, we are just beginning to develop a rudimentary scientific understanding. A complete picture of the photon and all its cousins, the family tree of quanta that make up physical reality as we now know it, has not been rendered by the current scientific paradigm. The knowledge that consciousness may be directly involved in the creation and perpetuation of the structure of reality forcefully turns our attention in a totally new direction. We must ask: Just what is this seemingly ineffable something we call consciousness, and exactly how is it linked to the matter and energy of the physical universe?

The cover of the current issue (July 1, 1995) of the weekly Science News displays a computer-generated picture of what looks like a garden spider's web in the moonlight. Only this artistic spider has woven some bright yellow, green, pink, and red strands, like fine yarn, into the silvery bug target. The caption reads: "Quark Hunt: Tip of the Top". It's a picture of a printout from the computers attached to Fermilab's

Tevatron collider, located west of Chicago, Illinois, documenting the first "hard" evidence of the existence of the top quark, the very last one to be detected, of the six quarks that are predicted by the standard model. The evidence is represented by one of the colored strings in the web, a tell-tale track leading away from the collision of a proton and an anti-proton.

Is this track evidence of the existence of a quantum particle in flight like a little missile, contradicting the Copenhagen interpretation? No, not at all. The "track" is not a continuous trail passing through all the points of space between the collision point and the edge of the spider web picture. Remember, the picture is only a computer-generated representation. If we were to look at the actual track with magnification, we would see a discontinuous line of little puffs of smoke-- actually ionized gas, where the quantum particle has smacked into something. Again, we are only seeing the effects of something quantum, not the quantum something itself.

All the structures of matter, and thus all of the physical properties of the elements, molecules, compounds, crystals, and cells, right up to and including the structures we actually see in the world of our everyday experience, have been traced to combinations of 12 elementary particles: six quarks and six leptons. All 200 plus particles that have been described over the years since Rutherford revolutionized target practice by shooting at a thin metal foil with alpha particles (helium atoms stripped of their electrons) can be accounted for with combinations of quarks and leptons. Actually, most of the structure of the stuff of the world which we experience is made up of only two types of quarks and one type of lepton: the up quark, the down quark, and that very well-known lepton, the electron. Most of the other quarks and leptons are as rare and fleeting as the scruples of a lawyer who's caught short of funds when his jaguar payments are over due. Their effects are detected almost exclusively in particle accelerators like the Tevatron.

The Tevatron is an underground circular metal tube four-miles in circumference in which physicists accelerate quantum particles (even though, of course, they never see them) around the curved track at near light speed using powerful electromagnets. They are smacked into atoms and other particles to create particle shower effects for physicists to study. Numerous sophisticated electronic sensors surround the hollow metal doughnut and send millions of bits of information to computers. This data is stored and analyzed by complex algorithms designed to sort out the data generated by collisions and tracks of interest to physicists and turn them into spidery diagrams like the one depicted on the cover of Science News.

Looking at computer-generated pictures, or the equations describing subatomic reactions, or one thing is clear: these are symbolic representations, several times removed from the phenomena they purport to describe. They are the products of ingenious methods, contrived by conscious beings who are trying to probe and understand a reality they cannot see directly. But the Copenhagen interpretation tells us that quanta cannot be separated from the apparatus by which their effects are observed. By logical extension, we have proved that they likewise cannot be separated from the sensing organs or the consciousness of the observer. In short, reality is not a conglomeration of unrelated objects, it is not fragmented, it is whole. It is only one thing. The illusion of separate objects is created by individual acts of observation. The limitations of the apparatus of observation, including the eyes of the observer, bring about the selection of specific, limited aspects from the whole of reality to make up the pictures of reality that we perceive through our senses. The illusion is that we tend to think that our little pictures tell the whole story.

In chapter 2 we proved that something real, but non-physical (that is, not composed of quanta), which we called consciousness, must exist, if we are to avoid the contradiction of infinite descent, to serve as the final receptor in the chain of phenomena reaching from the elementary particle being observed to the conscious entity doing the observing. But just how far can the conscious receptor be separated from the quanta of the observation, if they are both part of the same thing, and it is the act of observation that causes them to appear to be separate?

Let's look at the chain of events again. An observer, let's say an experimental scientist, believing himself to be separate from the rest of reality-- as most of us do, seeks to understand the nature of physical reality. He starts by isolating smaller and smaller bits of matter and attempting to study their structure. He eventually succeeds in isolating a few atoms in a Wilson cloud chamber. Then he bombards them with alpha particles, and finally he observes the effects of a quantum particle, a track of ionized particles. Of course, he doesn't actually see the quantum particle, or even the ionized particles. He only sees the effects of a series of quantum particles and receptors. The following list represents the least number of quantum and receptor steps that are involved in his observation:

1.) The quantum particle that is the object of the observation
2.) The quantum particles effected in the material registering the effects of the particle being observed, e.g., ionized gases composing a "track" or the mark of a photon's effect on a photographic plate, etc.
3.) The photons that carry the information of the effects registered on the first receptor to the retina of the observer's eye
4.) The elementary particles that make up the light-sensitive cells of the retina
5.) The electrons that carry the impulse through nerve tissue from eye to brain
6.) The quantum receptors in the brain
7.) The non-quantum receptor of consciousness

Since an observer becomes aware of the physical form of any object of observation through the effects of quanta of matter and energy, then, counting from the quanta of the object, the non-quantum receptor of the consciousness of the observer is at least the seventh step. Of course, with most observations, like that of the top quark evidence, many more steps are involved. The seven steps described above are involved even in the observation of the picture of the computer-generated diagram. In the top quark detection, the components of the particle sensors, the electrical circuitry, the electrical current carrying the bits of information to the computer, the computer, the ink and paper of the printout, and any other physical components that happen to be in the chain from the quark of interest to the consciousness of the observer, all are made of quanta.

It is interesting to note that the same reasoning applies to the perception of external quanta through any of the senses. The sound made by clapping hands, for example, originates from the interaction of quanta of matter and energy in one hand (1.), colliding with, and thus creating an effect on the quanta in the other hand (2.). The sound is then carried by molecules of air (3.), to the ear drum (4.), by the audio nerve (5.), to the brain (6.), and finally to the non-quantum receptor (7.) of a conscious observer (or, in this case, a conscious listener).

And so, it seems that the objects of the external world and the consciousness of the observer are separated by at least five layers of quantum effects. However, if all of these quantum systems are necessary to convey the evidence of the observed particle or conglomeration of particles to the awareness of the observer, then the last system of quanta, and *only* the last system of quanta, must be in direct contact with the consciousness of the observer. Either the non-quantum receptor does not expand to contact other quanta along the chain, or only the structure of the matter and energy related to the final set of quanta is specialized in a way that allows the transfer of information from the last quantum of matter or energy to the non-quantum conscious receptor.

This brings us to the question of the nature of the beyond-the-quanta receptor of the conscious observer. And here, we find ourselves in what, at least to physical science, is truly virgin territory. So far, most physical scientists have not even recognized the *possibility* of the existence of something real that can receive impressions from matter and energy, that is not itself composed of matter and energy. In this book, however, we have demonstrated that such a substance must exist.

It is not uncommon, when investigating virgin territory, to find that new concepts require new terminology. We need to find an appropriate term that will allow us to move beyond the awkwardness of the use of such terms as beyond-the-quantum or non-quantum receptor. A discussion of the functions of this new element of reality may suggest a simpler way to describe it. As the necessary bridge between matter and consciousness, it is a link, or *nexus*. It is the receptor of the information carried by the final set of quanta in the chain leading from the object of observation, and so it might be called a reception chamber or *receptorium*. Beyond the interface, images of reality are formed. We might call this formation space the *gestaltenraum*.

What are the attributes of the final receptor? It has to be capable of receiving bits of information from the physical properties of quanta such as charge, spin, velocity, and sequence, and be capable of assembling and structuring this information in such a manner that it creates and supports a consistent visualization of the external world. Through the consistency of experience, we come to believe that the details of the images formed here have a one-to-one correspondence to an external objective world.

We proved in Chapter 2 that this receptor cannot be composed of quanta. If it is not composed of discrete units, it must be continuous, or in Bell's theorem terminology, nonlocal. But we know that it does not pervade all space, since, if it did, no chain of particles would be necessary to relay information from the physical world. Therefore, it must be a bounded nonlocality, continuous only within a specific volume of space.

Light, as we have said, is a complex phenomenon. Considered as localized physical objects, i.e., photons, it appears to obey rules of its own, rules that don't seem to apply to other physical objects. It is always seen to travel at one speed, **c**, very nearly 300,000,000 meters per second (about 186,292 miles per second) by any observer, regardless of relative motion. Most moving objects whose speed we measure, like automobiles, baseballs, bullets, even rockets and space vehicles, obey the addition of velocity vectors rule.

The relative speed of two cars travelling toward each other, for example, is the sum of their speeds relative to a stationary observer. Light doesn't behave this way. The photons from two flashlights aimed at each other do not meet of a combined speed of 2 **c**. The ordinary addition of velocity vectors has to be amended for relativistic effects.

In fact, it is actually the other way around. In spite of its common-sense applicability, the addition of velocity vectors is only an approximation. Nothing really obeys the simple addition of velocity vectors law. Every observer has his or her perception of time and space affected by relative motion, it's just that, unless the relative motion is large relative to the speed of light, the time and space in the vicinity of observers in relative motion are nearly identical with time and space in the case where the observer and the observed are at rest relative to each other.

Einstein assumed that photons were localized objects moving through space at the speed **c**, and allowed time and space itself to adjust with relative velocity to assure that the relative speed between two objects never exceeds **c**, light would always be found to travel at the speed **c**, conforming to actual experimental measurements. But, just like Planck's constant, **h**, there is no explanation for why light travels at a constant speed, or why it has the specific value, **c**, that it does.

With our new understanding of the relation of consciousness and observation to physical reality, we are now in a position to understand and explain the how and why of the observation of light at the constant speed **c**. Speed is a pure measure of space and time. If an object traverses 100 miles (space) in one hour (time), it makes no difference whether the object is a baseball or a Mack truck, its speed is 100 miles per hour. And light is the medium of observation. Without observation, no quanta exist, no Mack truck, no baseball. Without quanta, there is no time and there is no space. Therefore, time and space are concepts created by observation.

There is a curious similarity between understanding the seven levels of quantum sources and receptors and the following Biblical passage.

> And when the seven thunders [transmissions of information from quantum sources to quantum receptors] had uttered their voices, I was about to write: and I heard a voice from heaven saying unto me, Seal up those things which the seven thunders uttered, and write them not.
>
> And the angel which I saw stand upon the sea and upon the earth lifted up his hand to heaven.
>
> And sware by him that livith for ever and ever, who created the heaven, and the things that therein are, and the sea, and the things that are therein, that there should be time [and space?]no longer.

> But in the days of the voice of the seventh angel, when he shall begin to sound, as he hath declared to his servants the prophets.

- The Book of The Revelation of St. John the Divine, 10: 4 through 7.

Without the seven levels, from the object of the senses to the receptorium, "time should be no more". Perhaps when we know the nature (i.e., hear the voice) of the seventh level (angel), we will have a better understanding of all of reality, including matter, energy, time, space, individual consciousness, and primary consciousness.

Clearly, **c**, the maximum speed obtainable by a quantum particle and **h**, Planck's smallest unit of action, owe their unique numerical values to, and can be derived from, limitations due to the relative size of the smallest possible units (quanta) of mass and energy. Thus the universal constants of relativity and quantum mechanics, and all such constants, are related to the physical limitations of the apparatus of observation, ultimately the sense organs. A parallel universe with the apparatus of observation constructed from quanta of a greatly different size, larger or smaller, would not be visible or detectable through the use of our sense organs.

Since light is the medium of our perceptions, understanding the nature of light is one of the most important keys to understanding the nature of the universe we perceive. Einstein discovered the principles of relativity by imagining that he was running along side a beam of light and observing it. We know now that consciousness is not separate from that which it perceives, so we can go Dr. Einstein one farther: We can imagine that we are the light.

Imagine that you are a photon traveling from a light source to a receptor. You are travelling at the speed of light relative to everything in the space through which you pass. What would you see? Applying the time and space transformation equations of the theory of relativity, we find that the lateral dimensions of everything you pass must be zero, since from your point of view, they are passing you at the speed of light. And time will have slowed to an absolute stop. In other words, from your point of view, there is no space or time in the world around you. You are simultaneously at the source and the receptor, in an eternal present, now, with no past or future. Reality, no longer separated by conscious observation has returned to a state of oneness where all things are in immediate contact.

Which is real, the severed and divided state that we perceive through the senses, or the unified reality of the photon? Obviously, what we consider to be the normal experience of the world is what we have to work with, for the most part. However, we have seen how science, operating with the assumptions of separation, produced contradictions that can only be resolved by reversing those assumptions. We saw how analyzing matter and energy as separate from, and unaffected by the consciousness of the observer, led to the opposite conclusion. Realization of the underlying unity of reality should give us the basis for a new paradigm capable of integrating our search understanding.

Now that we have proved that the non-quantum receptorium exists in every conscious observer, we might ask where in the physical structure of the observer it is located. Biologists and neurophysiologists have been asking a similar question in their search for the seat of the functioning of consciousness. But most of these scientists are starting from the assumption of functionalism and the theory of emergence. In other words, they are trying to explain consciousness as an abstract function emerging from the evolution of matter at a certain level of complexity. Could it be that the reality is exactly the opposite, that the complex structures of matter that allow the existence of conscious observers are the result of the functioning of primary consciousness to bring about the flow of order and structure into physical reality?

One researcher who has located what may be the principle area of the brain where quanta are in contact with the receptorium is the Nobel Prize winning British neurophysiologist Sir John Eccles. Eccles proposes that quantum-sized synaptic junctions, which he calls microsites, may be activated by mental patterns or informational codes in an area in the brain, near the top of the head, called the supplementary motor area, or SMA. (See The Wonder of Being Human, Sir John Eccles and Daniel N. Robinson, Macmillan, New York, 1984.) Even more remarkable is the fact that he also proposes that this area is the interface between the brain and non-physical consciousness. This position places Sir John in a very small minority of scientists who are categorized philosophically as dualists.

The biochemical evidence for Eccles' point of view, substantiated by the research of Kornhuber and Deecke, Lassen and Roland, and others, is very interesting: Experimental tests show that when a subject performs an intentional act, such as moving his or her fingers in a specific sequence, synaptic firing occurs in the SMA before the movement starts, and synaptic firing also occurs in the SMA when the subject merely thinks about performing the movements. Eccles concludes that such SMA activity is initiated by mental intention and not by any physical cause.

With the proof of the existence of the non-physical receptor presented in this book, the question of the physical location of the nexus between consciousness and the brain, and the question of whether the size and functioning of this nexus is fixed, or may be expanded in some way, continue to be important topics for further research. For the purposes of the discussions and considerations in this book, we will turn away from the conventional probing of matter in favor of investigating the form and structure of consciousness and its relationship to matter. Having uncovered the functioning of consciousness in the form of a real, non-quantum substance in contact with a part of the physical brain, we may proceed to the question raised at the beginning of Chapter 2: What happens in consciousness when an observation is made?

The information carried by the last in the chain of quanta, the first of which originated in the object of observation, is received in the receptorium and used by individualized consciousness to construct images in the gestaltenraum. The gestaltenraum, however, does not have to be a space located in the physical brain, because the images it creates are not necessarily constructed of quanta in physical space-time, they may be constructed in the nonlocal space-time of consciousness.

The mental space-time that exists in consciousness is far more flexible than physical space-time, because it is non-quantized. The infinitely divisible continuity of the real number line of mathematics, for example, originates in consciousness, not in physical reality. In fact, based on what we have just discussed, it can be argued that all structure and order has to originate in consciousness, not in the physical universe, because, as the second law of thermodynamics tells us, physical systems, left to themselves, will degenerate to the maximum entropy and equilibrium of total disorder. Where then, does the order and structure come from? It had to be latent or inherent in primary consciousness. It had to be selected from the multi-dimensional continuum of possibilities by the first receptor which was proved necessary in chapter 2.

At this point, we begin to see that reality, in the broadest sense, is actually composed of the substance of primary consciousness, that matter, energy, time, and space are observable features of this substance, comprising an infinitely continuous, all-encompassing, nonlocal whole. The illusion of separateness, is created by the initial drawing of a distinction, allowing an interplay between the separated parts, creating an ever-increasingly complex structure of localized quanta, as more and more of the innate patterns of primary consciousness are externalized. As they grow in complexity and sophistication, the externalized forms become more complete reflections or manifestations of the innate structure of primary consciousness, increasing their scope of consciousness and awareness. At some point in this process of externalization, the externalized forms, as conscious, but limited versions of primary consciousness, are also able to draw distinctions and make observations. These observations, however, are limited by the finite structure of the quanta making up the apparatus of observation with which the externalized or individualized consciousness is associated.

In this scheme of things, the individualized conscious entity, whether a hummingbird, dog, or human being, is only aware of that part of reality that his or her specialized sense organs make available. The sense organs of the conscious observer select the effects of specific parts of the infinitely continuous whole of reality for observation. The effects of those specific parts make up the impressions of reality that we receive through the senses.

Light has no objective existence until it registers on a receptor, and that receptor could not have come into existence without the action of consciousness. Without observations, the whole of reality remains like the world of the photon before it finds a receptor, nonlocal and timeless. The primary state of consciousness may consist of endless patterns which remain there, dreamlike, until, through the action of the drawing of a distinction, a boundary between light and dark, self and other, the archetypical patterns of primary consciousness are set free in the form of potential elementary quanta. And these potential elementary quanta may, if selected by observation, become photons which appear to fly like arrows, seeking targets to form and sustain the universe it as we now see it.

Through the filter of the physical sense organs, evolved for that purpose, the non-quantum substance of individualized consciousness becomes the target, receiving the effects of quanta that seem to fly like arrows to create the images of a world of shadow and light, pleasure and pain, peace and conflict. But we've learned that there are no arrows, only effects, selected by the senses and organized by consciousness.

Once the Persian King, Harun Al-Rashid, made Abu Vahib ibn Kufi, whose nickname was Bohlul (Great Laughter) accompany him into battle. Bohlul took a bow, but no arrows. When questioned by the king, he said:

"Arrows will come from the enemy."
"But, what if they don't shoot any arrows?"
"Then it won't be much of a battle."

- A Sufi story

Adapted from: Another Way of Laughter, a collection of Sufi humor, by Massud Farzan, E.P. Dutton & Co., Inc., New York, 1973.

CHAPTER 6.
CONSCIOUSNESS -
THE FAMILIAR UNKNOWN

The Void needs no reliance.
In space shapes and colors form,
But neither by black nor white is space tinged.
From the Self-mind all things emerge,
The mind by virtues and vices is not stained.

The darkness of ages
Cannot shroud the glowing sun

Transient is this world,
Like phantoms and dreams, substance has it none.

Renounce it and forsake your kin,
Cut the strings of lust and hatred,
And meditate in woods and mountains.

- Excerpts fromTilopa's Song of Mahamudra,
composed about 990 A.D. in Tibet

In 1957 I was a physics-math major at Central Methodist College in Fayette, Missouri. My roommate was a pre-theology English-philosophy major. Later, he earned a Ph.D. in seismology and became widely known as the foremost authority on the New Madrid Earthquake Zone, and I became a research hydrologist with the U.S. Geological Survey, and later, an environmental engineer and hydrogeologist. At the time, however, we were both very much interested in the relationship of mind to matter and the young science of parapsychology, a subject provoking considerable interest and controversy, due in large part to the publication of interesting accounts like the Bridey Murphy story, and the work of Dr. J.B. Rhine, at Duke University.

With the unpublicized support of the heads of the Physics and English Departments at Central, (They didn't want to lose their jobs.) we conducted a series of very successful experiments, one of which was recorded in the files of the Parapsychology Department at Duke University. Dr. Rhine encouraged us to pursue our interest in parapsychology, but cautioned us not to fail to obtain some training and skills in a field that would enable us to make a living. We both eventually took the advice to heart. But that is another story.

In one of the experiments at Central College, we were able, under controlled conditions, to receive perfectly a detailed series instructions that were written on a clipboard pad by a third, skeptical party, out of our sight and hearing, in another room. Because of the detail and sequencing of the instructions, the probability of our having reproduced the instructions by chance were virtually nil. In other experiments, we were able to obtain detailed verifiable information that neither one of us could possibly have known.

Once, while home on vacation from school, I tried a variation of one of our experiments at home with my parents. My dad and I attempted to receive letters of the alphabet in our living room that were being written on a pad by my mother in the kitchen. I had asked her to write a series of five letters, block style. We received five letters. When we went into the kitchen to compare the results, we found that we had a perfect match! The same five letters, in the same sequence. The probability of this occurring by chance, even making generous allowances for two of the letters being letters with a high use frequency in the english language, is approximately one in 343,200. These experiences, and others over the years since, have convinced me that the direct contact of individualized consciousness with physical reality may, under certain circumstances, related to concentration, visualization, and belief, be extended beyond the normal everyday boundaries of the *gestaltraum* in the brain.

By proving the necessity of the role of consciousness in the precipitation of physical forms out of the "great smokey dragon" (an apt term credited to John Wheeler) of Schrödinger's wave equation probabilities, we have taken a giant step forward in our quest to understand the nature of reality and the body-mind relationship. Beginning with the materialistic assumptions of an Einsteinian or classical local reality, pursuing our study of the structure and nature of matter and energy as far as we are able, we find that the physical world ultimately reveals itself to be much more than particles and waves in time and space. We know now that reality cannot exist in the form that we perceive without the action of consciousness, that it shares or reflects the logical structure and the nonlocal nature of consciousness, and that time and space are artifacts of consciousness, illusions created by the formation of distinctions in and by consciousness. But there are still many unanswered questions, including the question of the nature of consciousness.

What is consciousness? We all have first-hand knowledge of consciousness. In fact, it is the only thing we really do know first hand; the only thing we experience directly. And perhaps therein lies a major part of the problem: as soon as we ask the question "What is consciousness?" perhaps we have already gone astray. We have already missed the point. Because consciousness is not a what. When you say: "I am conscious", what does it mean? What is the "I" of this statement? Is it the whole body, the organism that utters the sounds? Is it something more, or something less? Is it dependent upon the organism, or is the organism dependent upon it? Is it the result of your experiences since birth, or is it more? Is it composed of thought, or is it the source of thought? Does it originate with and in the brain, or does it simply use the brain and body to express itself in physical terms? There have been many answers offered for questions like these, but few of them are scientifically verifiable. Another part of the problem is that we instinctively think that we know a lot about consciousness. After all, it is the essence of our very being. It is difficult, however, to know that which is the very foundation of our being. Asking someone to define consciousness is like asking a fish to find the beginning of the ocean.

So how do we discover the most basic reality of consciousness? Maybe consciousness resides in brain cells and nerve tissue. But each and every cell is composed of matter and energy, ultimately particles and waves, --none of which exist until a conscious observation is made. Because your awareness of the universe is facilitated by the organism you call your body, you tend to believe that somewhere within the body is where your consciousness resides. But if we remove one molecule or atom at a time, it is very difficult to say at what point your consciousness ceases to exist, --or, even _if_ it ceases to exist. We know that when portions of thc brain are cut away, or accidentally destroyed, other portions may take up the functions of the lost brain cells. When a sense organ has been destroyed through accident or disease, another sense organ or part of the body has been known to at least partially perform the functions of the lost organ.

It is much easier to say what consciousness is not, than to say what it is. _Can_ the basic reality of consciousness be discovered by the process of elimination? Or will it be like peeling away the layers of an onion, leaving us with a handful of nothing? As soon as we identify a candidate for investigation as the seat of consciousness, it becomes an object of investigation, and therefore distinct from the consciousness that is investigating it. It is therefore obvious that consciousness can not be anything that is conceived of as an object, but this approach eliminates everything that we normally perceive and think of as objective reality. The receptorium, defined in the previous chapter, is non-physical and nonlocal, and therefore, not an object as the term is normally defined, but it is associated with certain physical structures in the brain. Is the receptorium consciousness itself, or just an "organ" of consciousness that interfaces with the ordinary quanta of matter and energy that make up the physical world?

No one has seen an electron; but we know electrons by their effects. Even though we may not know in detail exactly what an electron is, the assumption that electrons exist has allowed us to explain electricity, electronic effects, and to develop most of modern technology. The physical properties of the electron, so far as we know them, are revealed by the effects, such as charge, current, and forces, that are attributable to them. What are the effects attributable to consciousness? You may answer: intellectual activity, thought, art, music, culture, etc. But there is a more basic answer. When we think of consciousness, we tend to start somewhere in the middle of the story. This is because we are immersed in consciousness; we Exist in the middle of the story. We know the world of consciousness through thinking, through art, science and culture. But if we are to establish a real science of consciousness, we must be very objective and start at the beginning. What are the basic characteristics of all of the effects that may be attributed to consciousness? The effects brought about by the operation of consciousness exhibit structure, form, and order. In the terms of physics, negative entropy. Consciousness affects physical reality by drawing distinctions and organizing those distinctions into logical structures.

If we accept the fact that the real, measurable effects of consciousness operating in the physical universe are form, structure, and order, what are the characteristics or functions of primary and/or individual consciousness that allow them to establish and maintain stable structure and order in the universe?

1. The primary, most basic function of consciousness is the drawing of distinctions.
2. The secondary function of consciousness is to organize these distinctions into various forms and structures, thereby decreasing entropy.

Taking a totally objective look at consciousness, we soon realize that we knew all along, even before Bell's theorem and the Aspect experiment, that consciousness interacts with the material world. For example, a structure originating in an architect's mind is projected into the physical world through the activities of construction as real structure: a building or group of buildings. The assumption of absolute separation of consciousness and matter is only an artifice, a way of simplifying the task of probing reality. We know now that our individual consciousness has definite and direct effects upon the physical organisms we know as our bodies. Medical science is just beginning to accept the fact that an individual's over-all belief system and state of mind can have profound effects on physical health and material well being. And, as we have seen, there is also evidence that individual consciousness can have direct perceptions of, and perhaps even direct effects upon, physical reality outside the human body.

Quantum physics has produced the evidence that has forced us to realize that consciousness is a necessary ingredient in the formation of the elementary quanta of physical reality, and thus of all structure and order. The primary and secondary functions of consciousness, i.e., the drawing of distinctions and organizing them into structures and forms, are not exhibited by human beings alone. Animals, plants, and even minerals manifest a tremendous amount of structure and order. It was suggested earlier that consciousness has more than one form. If we recognize all structure as the result of the functioning of consciousness, it becomes obvious that consciousness must operate on many levels, and through many forms and structures. And so, we must expand our view of the scope of consciousness, from considering it to be primarily the functioning of individual consciousness within the human brain, to all forms of structure and order. In fact, there is no reason to suppose that the functioning of consciousness is limited to living organisms. Stanislav Grof, in his book <u>The Holotropic Mind</u> (Harper Collins, New York, 1992), has this to say on page 105:

> Experiences of this kind [transpersonal extensions of individual consciousness] suggest that there is a constant interplay between the inanimate objects we generally associate with

> the material world, the world of consciousness, and creative intelligence. Rather than being from two distinctly different realms with discrete boundaries, consciousness and matter are engaged in a constant dance, their interplay forming the entire fabric of existence. This is a notion that is being confirmed by research in modern physics, biology, thermodynamics, information and systems theory, and other branches of science. Observations of the transpersonal realm are beginning to suggest that consciousness is involved in the so-called material world in ways previously unimagined.

The question now becomes: Exactly how is consciousness involved in the "so-called material world"? How does consciousness bring order to the universe? We know now that consciousness functions not only on the organic level, but also at the quantum level. So let's begin by considering the role of consciousness in the first level of organic matter where quanta become the significant physical forms. This happens at the sub-atomic level where the quanta known as electrons play an important role in differentiating one element from another and in determining how atoms bond to form compounds and molecules.

Consider an atom from a cell in your brain. For the sake of argument, suppose that it is a key atom, in a key cell, that allows consciousness to function in your body. More than likely, it will be a carbon atom.

Carbon atoms, because of their stable and symmetric structure, are able to combine with hydrogen and oxygen atoms to form the organic compounds that support life as we know it on this planet. Carbon is made up of atoms with four electrons whirling around its nucleus. What makes this atom so stable? The reason that this atom, or, in fact, any atom persists long enough to support life is one of the mysteries still not completely answered by science. Until the advent of quantum mechanics, it was a complete mystery.

At the beginning of the twentieth century, Ernest Rutherford, a New Zealander, working in England, devised a way to probe the structure of the atom. As we mentioned before, he shot helium nuclei (alpha particles) like tiny bullets, through a thin metal foil. when one of the "bullets" hit a particle within the structure of the metal atoms it was deflected. By studying the deflection patterns from thousands of experiments, he was able to develop a picture of the arrangement of the electrons and protons within the metal atoms. The picture he developed was a surprise: The atom seemed to be something like a miniature solar system, mostly empty space, with charged electrons circling a dense nucleus.

It was well known that the atoms of different elements, when burned, produce light of different colors. Sodium, for example, burns with a bright yellow flame, potassium with a purple flame, etc. Niels Bohr saw the connection between these colors and the structure that Rutherford's experiments had revealed. The color produced by each element turned out to be composed of several colors. Bohr realized that each color, or wavelength within an atom's spectrum of colors corresponds to an electron orbit, giving each element its unique atomic spectrum related directly to its atomic structure.

As is often the case in science, the solution of one mystery gave rise to another, even deeper mystery: What causes electrons and protons to conform to specific patterns and structures? How did they get into that configuration, and why do they stay there? Bohr knew that the force holding electrons in their orbits could not be gravity. The force of gravity, equal to the product of the masses of the particles involved divided by the square of the distance between them, is far too weak to hold the high speed electrons in orbit and keep the atom from flying apart. The force holding electrons in their orbits had to be electrical charge. But it was well known that when electrically charged particles move in orbits, they radiate energy. A negatively charged electron, circling a positively charged nucleus should act as a tiny transmitter, sending out a portion of the atom's energy with each orbital trip. Since it only takes a fraction of a microsecond for an electron to complete millions of orbits around its nucleus, according to the laws of classical physics, atoms shouldn't even exist at all; their energy should be gone in a flash. If somehow, the big bang, thought to be the creation of the universe and everything within it, propelled tiny bits of matter and energy into these strange configurations, the resulting atoms should radiate their energy away in nothing flat. So the mystery was: what keeps the atoms, and thus the whole universe, in such stable, complex forms?

In an attempt to solve this problem and explain the stability of atoms, Bohr integrated what Rutherford had learned about atomic structure with Max Planck's discovery concerning radiating energy. From Planck's work with black-body radiation, Bohr knew that energy is always observed to radiate in multiples of a very, very small basic unit or quantum. Einstein had already worked out the relationship between the electrons in the outer shells of atoms and light quanta, known as photons, in his paper on the photoelectric effect. Bohr concluded that the reason the energy of an atom doesn't quickly radiate away, as the laws of classical physics say it should, is because atoms are made of quanta, and electron orbits are determined by quantum energy levels. In order for an electron to jump from one orbit to another, or to radiate away, the atom must either absorb or give up a quantum of energy.

Again, this answer simply moves us into yet a deeper mystery. Bohr's answer only tells us why atoms don't unravel in a matter of microseconds into radiant energy, not why quanta exist in multiples of discrete units, or how highly structured atoms come to exist in the first place. The following statement, reflecting our general lack of understanding regarding the existence of this basic constant of the physical world, is found on page 1 of one of the texts that was being used when the author took his first graduate course in quantum mechanics:

> All discontinuities in nature are meted out in units based directly upon h (6.625 x 10^{-27} erg sec ^{2}). The existence of this number and its particular size together form one of the greatest mysteries of nature.
>
> - C. W. Sherwin, Quantum Mechanics, Holt, Rinehart and Winston, New York, 1960

Independently, Werner Heisenberg and Erwin Schrödinger worked out two different ways to describe the observed details of the spectra characterizing atomic structure mathematically. Heisenberg used matrices, Schrödinger used the wave equation. Both mathematical forms fit the observations. But the surprise was that both models describe a sub-atomic reality that only exists as a spectrum of probabilities until an observation or measurement is made. Even Bohr couldn't quite believe what he had discovered. Surely, reality couldn't be so strange. How could anyone trained in the "hard" science of physics accept the idea that objective reality depends upon the functioning of consciousness? The very assumptions, language, and scientific culture that had nurtured these scientists equated objective reality with physical reality. As a result of this bias, Bohr eschewed what he saw as atomic mysticism and replaced the requirement of "observation" with the mechanical event of "registration": the point in time and space when the particle in question is brought out of the cloud of probabilities by "an irreversible act of amplification." Atomic Physics and Human Knowledge, N. Bohr, 1958.

Bohr invoked activities related to human consciousness to resolve the EPR paradox, but, as a philosophical materialist, he couldn't take the next step and investigate the possibility of the actual involvement of consciousness, and he didn't even attempt to explain the origin of atomic structure. He knew that it was incorrect to think that particles or waves existed and travelled physical pathways before they registered in collectors, counters, or on photographic plates. But he felt that they would surely register on those receptors, with or without conscious observation. Most physicists still hold this view, which they call "atomic realism". (Note the confusion of "real" with "material".) But, as we saw in the previous chapter, it is a belief, not a scientific conclusion, and it simply does not hold up under close investigation. To solve the mystery of the origin of the structure observed in quantum experiments, we must finally bring the element of consciousness into the study of physics, no matter how stubbornly it is resisted. We must investigate the functioning of consciousness in the act of observation, and determine how the activity of consciousness is related to the appearance of particles and waves.

Physicists and others who hold that, even without consciousness, the universe would still exist pretty much as we perceive it today, are either consciously or subconsciously trying to hold on to the assumptions of scientific materialism. They are ignoring what we have learned from relativity and quantum mechanics. Those who argue that there has been a physical universe evolving ever since the big bang, quite independent of consciousness, have forgotten what relativity taught us about time and space: There is no absolute time or space. Time and space can only be described from the point of view of an observer. Bell's theorem, the Aspect experiment, and the proof of the infinite descent of receptors, combined with the results of relativity and the two-slit experiments, demonstrate clearly that matter and energy can only be described from the point of view of an observer. Physics has no meaning without a conscious observer.

Bohr argued convincingly that sub-atomic particles do not exist apart from the apparatus with which they are observed, and that they cannot be said to have form until they register on the receptors. But the receptors are made up of the same structures of energy and particles as those being registered. Where did the structure of the experimental apparatus come from? It most certainly did not evolve without the involvement of conscious entities. And what of the structure of the human eye and the brain that consciously observes the experiment?

How does consciousness relate to atomic structure? Clearly, all the elementary units of matter and energy that make up an atom are composed of quanta. Is there a part of individual consciousness, your consciousness, that operates on the quantum level, creating and maintaining atomic structure, supporting the forms of material bodies, or does the primary form of consciousness continue to operate as a separate, pervasive form, originating and maintaining the stability of physical reality? The later seems more likely. Given the nonlocal nature of consciousness, however, these two aspects of consciousness may not be totally separate.

When I wake every morning the same walls are there that were there when I went to sleep the night before, distinguishing my bedroom from the rest of the house. But even these walls owe their appearance of permanence and solidity to structures that depend on distinctions that exist at the atomic and subatomic level. Human beings were not known to have made conscious distinctions at this level until Lord Rutherford began firing his miniature bullets, and yet, it appears that a world of human beings, animals, plants, and stars existed prior to Ernest Rutherford.

The Copenhagen interpretation, Bell's theorem, and the Aspect experiment have clearly established the fact that we exist in a participatory reality. One type of participatory reality that has been suggested is known as consensus reality. This concept proposes that reality as we experience it has been built up by the conscious thoughts and beliefs of all sentient beings and is constantly confirmed and maintained by the feedback of continuing observations. Objective reality, and even its past history, is thus subject to change, as the beliefs of participating observers change. Observations that tend to support this view include the fact that some experimental results seemto become easier to obtain as more scientists are convinced of the validity of the theory that would explain such results. However, the concept of consensus reality does not answer the question of first cause. How did the first form of sentient life arise with no prior consciousness to make the necessary distinctions and observations to bring the quanta of its physical form out of the spectrum of possible states?

This prompts us to turn to a slightly different form of participatory reality, in which consciousness consists of at least two forms: a primary form that precedes the formation of the physical universe, and a secondary form with a limited role, compared with that of the primary form, in the creation and maintenance of the physical forms of the universe. This two-levels-of-consciousness reality answers the question of first cause in a manner that is consistent with Bell's theorem, the Aspect results, and the proof of infinite descent. Like every other answer, it raises further interesting questions concerning the interaction of the two forms and their functioning in relation to each other and the physical universe, and it opens the door to a whole new world of scientific inquiry.

Returning to the question of the existence of the atom and atomic structure, we note that science currently has no explanation for the cause or functioning of the strong and weak forces, the forces that keep sub-atomic particles in place and govern particle decay. We know that gravity is a function of the masses involved and the distances between them. Electromagnetic forces are generated by relative motion. But what generates the strong and weak forces? These forces appear to be quantized energy fields associated with material quanta and thus related to the so-called universal constants: Planck's constant, **h**, and the speed of light, **c**. But what is the origin of these constants of observation? They are certainly related to the size of the quanta forming the molecules that make up experimental apparatus and the sense organs, such as the eye, retina, rods and cones, of the observer. The size of the quanta being observed must depend upon the size of the quanta of the means of observation. We have seen that Bohr's resolution of the EPR paradox, Bell's theorem, and the results of the Aspect experiment implicate consciousness in the objective manifestation of the forms of particles and waves. We should not be surprised to find consciousness intimately involved in the binding forces and universal constants governing the structural features of the universe. Could the relative size and rate of motion of the quanta making up objective reality, the observer and observed alike, all be dictated by features contained within the innate structure of primary consciousness?

What is the nature of consciousness that it can contain structure and project that structure into the forms of objective physical reality? The nonlocal and nonobjective aspects of reality anticipated by Bell's theorem and verified by the Aspect experiment are indicators of the involvement of consciousness; like signatures of consciousness in the physical universe. We can identify them as such by comparison with our experience of individual consciousness. The consciousness that we experience every day exhibits nonlocality. We are simultaneously aware of sensations at opposite extremes of our field of body awareness. For example, you may be consciously aware of tensing the muscles in your toes at the exact same instant that you are aware of placing a hat on your head. In the spaceless, timeless realm of thought, complex mental images of physically distant places like Paris and San Francisco or Hong Kong and Kansas City exist simultaneously in individual consciousness. And nonobjectivity is, by definition, subjective, i.e., contained in consciousness.

We must take care, however, not to equate subjective reality with fantasy, imagination, or hallucination. Objectivity can be applied to consciousness in the same way it can be applied to material objects. The term nonobjectivity used by John Bell may be something of a misnomer, but Bell can certainly be excused for using a confusing or misleading term, since he was referring to an aspect of reality hardly recognized by science at all, at that time. Once we accept consciousness as a *bona fide* part of objective reality, there is no reason to suppose that there cannot be objective structure existing in consciousness. Such structure is very much a part of reality, but not materially objective in the sense that it exists totally as an object among the objects currently regarded by the established scientific paradigm as physical reality.

As individual repositories of consciousness, encased in physical bodies, we are not generally aware of the fine distinctions of atomic and subatomic structure. Not many of us pursue the type of knowledge that Ernest Rutherford and Niels Bohr did. We are generally not directly aware of the forces that create and maintain the structure of the physical universe. However, it is entirely possible, and seems to be very likely, that the primary form of consciousness does include such direct awareness. Evidence of the operation of the primary consciousness, defying the second law of thermodynamics, is all around us.

By proper integration of the study of consciousness with the physical sciences, we will be able to discover and map the structure of primary consciousness. What we now know from relativity, quantum mechanics, and the study and experience of consciousness, suggests that when this structure is revealed, the origins of the universal constants, the strong and weak forces, and the integrating relationship of relativity and quantum mechanics will be known. Arising only imperfectly in individual consciousness, the laws governing all forms of reality will be found to comprise the natural structure of primary consciousness.

If we are successful in discovering and revealing the structure of primary consciousness, we will have the answer to how the structures of atoms and subatomic quanta are formed and sustained. If there was no physical universe until the big bang, the first conscious distinction projected in physical form had to be synonymous with the big bang. If so, all of the structure that has ever existed, or will ever exist, may be found to have existed in the primary form of consciousness prior to the formation of the first distinct particle by differentiation from the conscious background substance of reality. Just as the structure of a sculptured statue or a building exists in the consciousness of a sculptor or architect before it can be rendered in wood or stone, it seems very likely that the structure of quarks, atoms, molecules and planetary systems will be found to exist as innate structures within primary consciousness.

Given that there are two aspects of consciousness: primary and individual, and that they both participate in the formation of the reality we experience, just how do they relate to each other and to the perceptible universe? In order to attempt to answer these questions, let's return to Einstein's moon and the hummingbird outside my window. Does the moon exist when no one is looking at it? Yes, of course it does. Did it exist before there were human beings on the earth; before any life existed on the earth? The physical evidence suggests that it did. Does the hummingbird (and the human being) have a hand in the creation and/or sustenance of its own form? Undoubtedly we do. Knowing, as we do now, that quantum particles and waves do not emerge from the multiplicity of possible forms that exist within the structure of primary consciousness until they are registered as impacts on a receptor, the real question becomes: How did the particles and waves that make up the moon, the hummingbird,--or any other physical object, come into existence in the first place?

To answer this, we must take a hard look at the nature of reality as revealed by the two-slit, delayed-choice experiments, Bell's theorem and the Aspect experiment: We know now that reality is nonlocal, whole, and self-referential. The Copenhagen interpretation implies that a wholeness exists at the quantum level of reality and Bell's theorem defines this wholeness as nonlocality. We have shown that the action of primary consciousness is necessary for anything physical to exist. Thus the nonlocality of primary consciousness is universal. And if everything that exists is part of one continuous whole, then reality is ultimately self-referential. A number of physicists, logicians, and philosophers have reached this conclusion. G. Spencer Brown, for example, says on pages 104 - 106 of Laws of Form:

> Let us consider, for a moment, the world as described by the physicist. It consists of a number of fundamental particles which, if shot through their own space appear as waves, and are thus of the same laminated structure as pearls or onions, and other wave forms called electromagnetic which it is convenient, by Occam's razor to consider as travelling through space at a standard velocity. All these appear bound by certain natural laws which indicate the form of their relationship.

Now the physicist himself, who describes all this is, in his own account, himself constructed of it. He is, in short, made of a conglomeration of the very particles he describes, no more, no less, bound together by and obeying such laws as he himself has managed to find and to record.

Thus we cannot escape the fact that the world we know is constructed in order (and such a way as to be able) to see itself.

This is indeed amazing.

Not so much in view of what it sees, although this may appear fantastic enough, but in respect to the fact that it *can* see *at all*.

But in order to do so, evidently it must first cut itself up into at least one state that sees, and at least one state that is seen. In this severed and mutilated condition, whatever it sees is *only partially* itself...

Thus the world, when ever it appears as a physical univèrse, must always seem to us, its representatives, to be playing a kind of hide-and-seek with itself.

And John Archibald Wheeler, in At Home in the Universe, The American Institute of Physics, Woodbury, NY, 1994, on pages 290 & 291, under the heading
"The Building of 'Reality' This Participatory Universe":

What lies over the hill? What are we to project ahead out of the present landscape's (science's) two greatest strangenesses (the bounds of time and the quantum)? Of these one, the "bounds of time," argues for mutability, law without law, law built on the statistics of multitudinous chance events, events which --undergirding space and time --must themselves transcend the categories of space and time. What these primordial chance events are, however, it does not answer; it asks. Unasked and unwelcomed, the other strangeness, the quantum, gives us chance. In "elementary quantum phenomenon" nature makes an unpredictable reply to the sharp question put by apparatus. Is the "chance" seen in this reply primordial? As close to primordial as anything we know. Does this chance reach across space and time? Nowhere more clearly than in the delayed-choice experiment. Does it have building power? Each query of equipment plus reply of chance inescapably do build a new bit of what we call "reality." Then for the building of all law, "reality" and substance --if we are not to indulge in free invention, if we are to accept what lies before us --what choice do we have, but to say that in some way --yet to be discovered --they all must be built upon the statistics of billions upon billions of such acts of observer-participancy?

One cannot help but notice the consternation and bewilderment which physicists display upon confrontation with the observation-dependent reality that their own carefully "objective" methods, assuming reality to be independent of the observer, have wrought. The bewilderment is, however, easily dispelled by relinquishing the materialistic bias and recognizing the role of consciousness in the formation of the reality we perceive. Every observation is an instance of reality/primary consciousness observing itself. It is the completion of a self-referential loop. Structure and form, originating in primary consciousness, is projected as a spectrum of potentialities. The process is completed when one specific structure or form is selected by observation and confirmed again in the nonlocal space of consciousness.

While it is easier to study form and structure as it appears in the physical world, if it originates in consciousness, we must ultimately seek it in consciousness itsèlf. But before we proceed in this direction, it is interesting to see what physicists have found when using current analytical methods to study the only "physical" reality that lies beyond the quantum: pure space, or absolute vacuum, something difficult to achieve in the laboratory, but assumed to exist or at least to be very closely approximated in interstellar space.

Applying quantum mechanics and relativity in an effort to calculate the force of attraction that might exist between two uncharged metal plates because of the electromagnetic fields induced by the motions of atoms and elementary particles, H.B.G. Casimir, a Dutch physicist, was surprised to find that the force indicated by this analysis was proportional to the inverse of the distance that separated the plates raised to the fourth power. Relative to the complexity of his calculations, this result was amazingly simple. Casimir discussed this with Bohr who surmised, based on the form of the expression, that the force had more to do with what separated the two plates than with what was in them, and suggested that he turn his attention and analysis to that which separated the metal plates: empty space itself.

This approach may sound very strange to anyone not familiar with quantum physics, but the reason Bohr suggested it was that the quantum theory of light, called quantum electrodynamics, or QED, predicted that a perfect vacuum, after the removal of all matter, heat, and even light, would still be filled with energy. This energy was called the zero-point energy of the vacuum, and it has been very troublesome in theoretical physics since. Integrated over all space, its magnitude is infinite. How can nothing be filled with an infinite amount of energy?!

At any rate, when Casimir approached the problem from this perspective, he arrived at the same simple expression for the attractive force between the two metallic plates by a much shorter and more elegant derivation. Reference: <u>On the Attraction Between Two Perfectly Conducting Plates</u>, Proceedings of the

Koninklijke Nederlandse Akademie van Wetenschappen, Holland, 1948.

One of the most challenging problems of physics, from the very beginning, has been the problem of the operation of forces across empty space, known as the "action at a distance" problem. Maxwell's original concept of the field seemed to be the answer, at least for electromagnetic radiation. With the advent of quantum mechanics, however, even fields had to be quantized, since everything, including energy, is found to occur in multiples of discreet units based on Planck's constant. The development of quantum electrodynamics led to the concept of "messenger" particles of energy, transmitting the forces of attraction or repulsion across empty space at the speed of light, and, as we said, it also led to the discovery of the infinite energy of the void. Since this infinite reservoir of energy could not be tapped physically, this was not considered to be a serious problem, --at least not before Casimir's calculations. QED posited quantum fields and Casimir showed (and later verified by experiment) that empty space is filled with energy. Quantum physicists concluded that there is a structure associated with the vacuum of empty space!

This structured vacuum between or beyond quanta, sounds strangely similar to the nonlocal primary consciousness that we have identified as the repository of all structure. There is, however, one very significant difference: Within the current paradigm, the energy and structure of the void can only be investigated from one side, since the materialistic bias of the paradigm permits nothing that cannot be explained in terms of matter and energy constructed of units that are multiples of planck's mass, in space meted out in Planck lengths. In transcendental physics, on the other hand, we have traced the formation of the reality we observe to the gestaltraum, the nexus where consciousness and matter meet, the seat of individual consciousness where the loop of observation is completed when the structure of some selected aspect of primary consciousness is registered in individual consciousness. We can now begin to investigate from the side of consciousness.

But what is the nature of this link between matter and consciousness? We have called it a nexus, a receptorium, a gestaltraum. It must perform all of these functions. It is a connection between the ostensible physical world and the more subtle, nonlocal conscious realm that underlies and penetrates it. It receives information carried by electrons, and possibly other elementary quanta, and it interprets and assembles or forms that information into a meaningful image that enables the organism within which it functions to interact with limited aspects of reality. But these are functions that it has to perform, not what it is. It is not composed of quanta of matter or energy so it is like the empty space of Casimir's experiment, existing beyond, and possibly even within the structure of the quanta it receives. It exists at the interface of matter and consciousness, as a boundary for both, and perhaps, like a boundary it may occupy no specific volume of space.

How do we deal with such a thing? We know that it is real. The quantized information coming into the brain has to have a receptor in order to register. Otherwise, they would remain in the nondistinct spectrum of potentials represented by Schrödinger's wave equation. And the receptor has to be nonphysical to avoid the infinite descent of receptors, which is impossible with the finite quanta of the physical universe.

While puzzling over this, it may suddenly occur to us that there may be a way to proceed: When studying physical reality, like stars, planets, stones, and atoms, we use physical instruments and apparatus. The receptorium/gestaltraum has the same elusive nature as consciousness. Why not use conscious apparatus to look at it?

There are methods, developed mainly in the East where consciousness has not been so vigorously barred from the list of acceptable elements of objective reality, that lead to direct observation of some important aspects of consciousness. These observations are striking, meaningful in relation to the reality of both mind and matter, and they are repeatable and verifiable, virtues that we must require in any scientific investigation. And it can be said that technically, these results, direct observations of the objective form and structure of consciousness, are available to anyone; but not everyone will achieve valid results on the first, or even on the tenth attempt. Of course, the same can be said for theoretical physics.

Some of these methods, which we will discuss only briefly here, are common, to a number of well-established disciplines that are practiced by relatively small groups within the context of the world's major religions. These are group such as Zen and Tibetan Yoga in Buddhism, Raja Yoga in Hinduism, Sufism in Islam, and Judeo-Christian Mysticism. The methods involve sitting in an alert position, clearing the mind of distractions by using specific procedures and techniques, and focusing the attention on certain distinct features that appear in the objective background of consciousness. Such exercises in consciousness research can be practiced with just as much intellectual honesty, careful objectivity, accuracy, and control as any experiment in particle physics. And the results can be just as valid and enlightening concerning the objective nature of reality.

If we adopt the attitude that consciousness is a *real* feature of objective reality, in exactly the same way that the physical world is, then our own consciousness may be regarded as an instrument with which we may probe and study the conscious aspects of reality in the same way we use a microscope or particle accelerator to probe and study the physical aspects of reality. A physicist knows that to obtain valid experimental results, the apparatus being used must be constructed with sufficiently accuracy and refinement and must be in good working order. The same is true of the instrument of individual consciousness. For example, achieving mental clarity and one-pointed concentration is analogous to cleaning and focusing the lenses of a microscope or telescope, though not nearly as easy.

We will return in a later chapter to a more detailed discussion of techniques and disciplines for refining individual consciousness to the point where one may use it as an investigative tool, but for now, we will continue with the development of the more abstract method of probing the logical structure of consciousness using a new logical tool: the Calculus of Distinctions.

Is it possible to probe the structure of consciousness in manner similar to the way Ernest Rutherford probed physical structure? Can we demonstrate that an objective structure exists in consciousness without reference to structures perceived in the physical world? If so, the argument for the existence of a primary consciousness containing the cosmic blueprint of all forms would be made even more compelling. The next chapter and Appendix D are devoted to using the Calculus of Distinctions to probe and reveal the innate structure of consciousness.

This pure Mind, the source of everything,
Shines forever and on all with the brilliance of its own perfection.

But the people of the world do not awake to it,
Regarding only that which sees, hears, feels, and knows as mind.

Blinded by their own sight, hearing, feeling, and knowing,
They do not perceive the spiritual brilliance of the source substance.

- From The Zen Teachings of Huang Po, Translated by John Blofeld, Grove Press, Inc., New York, 1958.

CHAPTER 7.
THE STRUCTURE OF CONSCIOUSNESS

Things derive their being and nature by mutual dependence and are nothing in themselves.

-Nagarjuna, ca. 200 A.D.

While working in the Middle East, I met and became friends with an engineer from Iran who was a collector of Sufi stories. The Sufi mystics, whose beliefs and practices are generally considered unacceptably radical by orthodox Muslims, originated somewhere in the once-fabled land of Persia, where they developed poetry and story telling to levels both sublime and ridiculous. The hero of many a sufi story is a strange little man called Mulla (a religious title of respect) Nasrudin. The following story illustrates some human foibles related to searching.

Mulla Nasrudin was on his way to the corner bakery to buy a piece of bread for his evening meal, when he dropped his only coin on the cobblestone street. As the light was failing, he got down on his hands and knees to feel for the errant coin. As the day's light continued to fade and he still hadn't found the coin, he groped his way toward the lighted corner, feeling the cobblestones as he went. A blind man who was standing under the street light at the corner, heard him groping the cobblestones and asked what he was doing.

"I dropped a coin. It's all I have and I'm hungry," he replied.

"Where did you drop it?" the blind man asked.

"In front of the mosque."

"But that's a hundred paces away! Why are you looking here?"

"Because the light's better here," Mulla muttered. "Look, since you're blind, your sense of touch is probably much keener than mine. If you can find the coin, I'll share the bread with you."

The blind man quickly agreed, but instead of going up the street, he went into the bakery. A few minutes later, he came out carrying a lantern. He started up the street, lantern in one hand, white cane in the other, feeling his way along the curb.

"What is this?" the Mulla exclaimed. "Why do you need a lantern? I happen to know you're as blind as a turnip. Day and night are the same for you."

"Yes, that is true," the blind man agreed, "but the lantern is not for me, it's for blind people like you, who would trample me while I'm down on my hands and knees in the dark!"

Researchers in the field of neurophysiology trying to understand the mind-body interaction are somewhat like the Mulla, looking for the answer in the wrong place. The Mulla did find part of the answer to his problem under the street light at the corner, but the real object of his search was not there. Similarly, part of the answer to the mind-body enigma is to be found in the structure of the brain, but the real coin is to be found in the understanding of the nature of consciousness.

Does primary consciousness have form and structure? If it does, it should be reflected, at least partially, in individual consciousness, and in the structure of the physical universe. If all structure is projected from primary consciousness, we can understand why logical structure, thought to originate in human minds, is repeatedly found to apply to the structure of the physical universe. The human mind and the physical universe display the same symmetry and logic because that symmetry and logic underlie all reality in the substance of primary consciousness.

If the universe is a reflection of logical order and structure, then perhaps there are no accidents. My discovery of the tool that reveals the structure of consciousness was the result of a string of seemingly random events that began one Sunday in 1984, when my wife and I had an early afternoon lunch at the Inn of the Seventh Ray in Topanga Canyon, California. After a wonderful meal served at a table on the flagstone patio under the trees, we strolled over to the bookstore behind this one-of-a-kind restaurant. Jacqui picked a book off a shelf, glanced at it briefly, and handed it to me. On a white dust jacket, decorated with nothing more than two nested, clean, straight-lined right angles, I read, in large black print: "LAWS OF FORM". Under that, "G. Spencer Brown". And yet below the author's name: "In this book G. Spencer Brown has succeeded in doing what, in Mathematics, is very rare indeed. He has revealed a new calculus, of great power and simplicity..." This bold statement, sure to grab the attention of any serious student of mathematics, was attributed to none other than Bertrand Russell, one of the most prolific writers and thinkers of the twentieth century.

"I think you need to read this book", Jacqui suggested.

I turned it over and read on the back: "...this book is surely the most wonderful contribution to Western philosophy since Wittgenstein's *Tractus*." -- Alan Watts

"I think you're right!" I agreed. We bought it and I could hardly put it down until I had read it cover to cover. I can't claim to have understood it all at the first reading, and I didn't realize the full significance of it, or how it related to the work that I was doing, until more than two years later. When that realization came, it was sudden and unexpected. It happened in Jeddah, Saudi Arabia in January of 1986. The way it happened was unique in my experience. Nothing quite like it has happened to me before or since.

I was working for an environmental consulting firm that had several contracts with the Saudi Government. My wife, nine-year-old son, and I lived in a British-owned and operated compound called Arabian Homes, in the As-Salamah District of Jeddah. At one-eleven a.m., I woke suddenly and sat up in bed. I had had a vivid dream. It seemed that a pebble had been dropped into a pool of dark, clear water. The pool had no bottom or surface that I could see. The resulting ripples spread through the water as if they were ever-expanding spheres of golden light. Smaller spheres formed within the original spheres, and many patterns of amazing complexity and dazzling beauty came into being as if by magic. This vision would have been memorable simply on the basis of its vibrant color and unusually striking visual clarity, but somehow I knew that I was seeing an archetypal form, a form virtually underlying the structures of universe. This form, I thought, was somehow related to the form of the most elementary particle, or perhaps even the first cause in the creation of the universe.

I got up and tried to write down a description of the experience. My thoughts about what I had seen made me think of Laws of Form. When I got the book out and opened it, the first line on the first page jumped out at me: "The theme of this book is that a universe comes into being when a space is severed or taken apart." As I leafed through the book, I found that somehow, I had a much deeper understanding of it now. I sat up and wrote through the rest of the night, and I continued writing for several days, almost without stopping. The result was the forerunner of the present book.

The universe we experience exhibits a variety of forms and structures. How were these forms created? Where does the structure of the universe come from? G. Spencer Brown's statement quoted above implies that the structure of an entire universe follows from the severing or taking apart of a space. How this may be so is not immediately apparent. In the Laws of Form, Brown describes the drawing of a distinction in a previously and otherwise unmarked space. If we take this space to be the pre-creation void, then the processes described by Brown's laws are the processes of the creation of structure and order in the universe.

In the previous chapter, we defined the two primary functions of consciousness as the drawing of distinctions and the organization of distinctions into form, structure, and order. Consciousness first draws a distinction between itself and other, severing the space of awareness, and then draws distinctions in the other, creating structure and order in the form of quanta of matter and energy. How does consciousness do this? In order to discover exactly how consciousness functions and participates in the creation and perpetuation of the physical universe, we must first understand the processes and consequences of the conscious act of the drawing of distinctions.

We are aware of numerous types of distinctions in the physical universe: distinctions of color, size, shape, and density; distinctions between matter and energy; distinctions of location, motion, and velocity. In the past, we have thought of such distinctions as having objective existence independent of conscious observation, but now we know that they do not. No elementary particle can come into existence without a receptor, and as we proved in the chapter 2, the final receptor must be an organizing factor beyond quanta. We know this factor as consciousness. To probe the innate structure of consciousness, we must forego external distinctions, temporarily leave the sights and sounds of the physical world behind, and objectively investigate the process and consequences of the drawing of a distinction without reference to any specific type of distinction outside of consciousness. Fortunately, this turns out to be less difficult than it sounds.

Much as cosmologists have endeavored to trace the history of the expanding physical universe back to the first few microseconds after the big bang, we may trace the progression of distinctions from those made consciously by individual consciousness every day, back to the original distinction that brought the entire universe into being. And just as Lord Rutherford isolated helium nuclei and fired them into the atomic structure of physical matter to observe the results, we may logically isolate a primary act of distinction, allow it to move through the region of consciousness in which it was formed, and observe the results in an objective manner. But first, we must find a direct, efficient, and logical way to represent the drawing of a distinction, and develop the mathematical tools necessary to describe the fundamental operations involved.

The derivation and development of the Calculus of Distinctions, devised for the specific purpose of describing the conscious act of the drawing of distinctions and the organization of those distinctions into form and structure, is presented in more detail in Appendix D. Recognizing the fact that not every reader is a mathematician, or even wants to be bothered with the logical symbolism, I've made every effort to keep mathematical notation out of the text as much as possible. But before we are ready to proceed with the investigation of how individual consciousness participates in reality and to discuss some of the more interesting aspects of transcendental physics, the reader needs to understand why this new mathematical language is necessary, how it is developed and applied, and what it reveals.

Most of the basic mathematical tools currently in use in the physical sciences, including numerical analysis, geometry, trigonometry, integral and differential calculus, etc., are based on the assumption of the absolute separation of the observing consciousness from that which is being observed. In other words, our basic mathematical tools are grounded in local reality and are therefore appropriate for describing the type of reality posited by scientific materialism. Knowing, as we do now, thanks to Bell's theorem and the Aspect experiment, that reality has both local and nonlocal features, we must develop a mathematical system capable of encompassing both. Such a system will not negate the existing disciplines of mathematics, but will supplement, complement, and include their root concepts as a subset. Being more fundamental than conventional mathematics, this system will be nonnumerical, and it will include the logical precursors of the fundamental operations of numerical arithmetic.

Before we begin to develop the mathematical language needed to describe the interaction of consciousness and matter, i.e., the mathematical language of transcendental physics, we must develop a basic understanding of the functioning of consciousness. Primary consciousness, for reasons that may become clearer as we proceed, has projected receptors which make selective distinctions or patterns of distinctions, perceived by physicists as elementary particles, in the infinitely continuous substance of the void.

Individual consciousness, in the form with which we are most familiar, operates through the organs and senses of a physical body to provide primary consciousness with a means by which to observe selected structures in physical form. The forms that we normally perceive are complex structures made up of literally billions of individual distinctions. These forms exhibit almost endless variety, and the individual distinctions of which they are built are so small, relative to the size of the cells comprising our sense organs, that we have to deduce their existence from the indirect evidence of their effects. Because of this disparity in terms of scale, between a single distinction and the forms we must deal with in the course of our day-to-day lives, the basic nature and origin of the phenomena we perceive are hidden from us.

The sense organs, nervous system, and brain act as filters, selectively reducing the number of distinctions that reach the receptorium. The limited nonlocal aspect of individual consciousness then acts as a gestaltraum assembling the bits of information received into patterns which we perceive mentally as specific aspects of the real world. The patterns thus assembled in individual consciousness reflect patterns inherent in primary consciousness somewhat like the manner in which a whole picture is reproduced even in a small part of a holographically imprinted film.

In order to comprehend the drawing of distinctions, the transferring of logical forms to objective structure, and the local and nonlocal nature of reality, it will prove to be helpful to consider distinctions from three perspectives: 1) within individual consciousness, 2) in primary consciousness, and 3) in physical reality. Individual consciousness both recognizes existing distinctions and creates new distinctions. Within individual consciousness, it is possible to conceive of idealized distinctions such as the basic geometric concepts of point, line, and plane. These distinctions are partial forms, limited to zero, one, and two dimensions, abstracted from real three dimensional forms. They have no existence of their own, apart from more complex forms of three or more dimensions. Because of the ability of individual consciousness to conceive of forms internally and perceive forms externally, forms that arise in individual consciousness may be either conceptual or perceptual. Since all forms having existence in external reality depend upon the action of consciousness, specifically the drawing of distinctions, they must also have existence in primary consciousness. Furthermore, since zero-, one-, and two-dimensional forms are conceptual forms abstracted from forms of three or more dimensions, we shall define a real or existential distinction as one that must exist in primary consciousness and in at least three physical dimensions.

We know that the structure and form that has been projected by primary consciousness into physical reality exists in at least four dimensions, three of spatial extension and one of temporal extension. We also know that all structure and form can be described in terms of distinctions. A Calculus of Distinctions must therefore be developed as a logical system capable of describing the manner in which the forms of reality are brought out of the multi-dimensional manifold of primary consciousness by the act of the drawing of distinctions. It must begin with the simplest act of consciousness, the drawing of a single distinction. To be successful, the Calculus of Distinctions must demonstrate how the forms of reality are systematically built up and constructed of various distinctions through the application of logical transformations. Such logical transformations may be expressed in the form of mathematical operations.

Mathematical transformations are accomplished through the process of calculation. Mathematical calculation is defined as the logical process of using one or more fundamental operations to transform an expression representing a given value into a new and different expression of the same value. (For example, $1 + 1 = 2$.) The drawing of a distinction qualifies as mathematical calculation since it involves the fundamental operation of dividing the universe or field of awareness into two parts: the part distinguished, and the rest of the universe. The value before the operation is the whole universe of awareness. The value after is the same, but that value is now structured into two parts, logically divided by the distinction that has been drawn. The drawing of a distinction is actually the most fundamental of mathematical operations, but it is skipped over and ignored in standard mathematics texts, because it is so basic. It is more basic than the so-called fundamental operations of the traditional system of mathematics, since it is more primitive, and actually precedes them logically. The Fundamental mathematical operations of addition, subtraction, multiplication and division depend upon the prior development of the concepts of equivalence and enumeration. And equivalence and enumeration depend upon the even more basic operation of the drawing of distinctions.

People currently working in science and mathematics are generally trained in the theory and application of the existing system and therefore, may find it difficult to absorb and embrace a new system of mathematical logic. The idea of a new mathematical language, capable of dealing with nonlocality and the functioning of consciousness, may conjure up visions of a highly abstract system of symbolic logic, difficult to understand and difficult to apply. But, in fact, as we shall see, the nonnumerical Calculus of Distinctions is simpler, more direct, and more descriptive of the reality we experience than the existing, more familiar forms of mathematics. Some problems and derivations that are difficult in conventional mathematics, are rendered so simple in the Calculus of Distinctions as to be almost trivial.

If there is difficulty, it lies not in the level of abstraction or sophistication involved in the Calculus of Distinctions, but, in fact, just the opposite: We are constantly drawing distinctions, perhaps hundreds every minute, some with deliberate intent, many of them involuntarily by the psycho-physical mechanism of conditioned response. The functioning of consciousness in the act of the drawing of a distinction is so close to us, so much a part of the way we function as sentient beings, that we simply overlook it.

While setting up the mathematics of the drawing of distinctions, we want to be sure that the structure it reveals is not simply a reflection of forms we have become conditioned to regard as existing "out there", independent of the functioning of consciousness. Toward this end, a subtle distinction must be borne in mind: the symbolism we use must not be taken to represent a distinction existing in physical reality, but the conscious act of the drawing of a distinction. We do not, therefore choose a circle drawn on a plane, or a sphere in three-dimensions as our basic symbol, since such symbols are related to forms that we are accustomed to regarding as already existing in a reality assumed to be distinct from ourselves. However, the choice of a symbol is arbitrary as long as it doesn't confuse the act of the drawing of a distinction with some objective physical form, and is capable of functional expression on paper.

For a number of reasons that are beyond the scope of this discussion and that only become apparent as the Calculus of Distinctions is applied to various logical and practical problems, it is convenient to choose the symbol ⌉, which is similar in form to that chosen by the English logician George Spencer Brown in Laws of

Form. (Brown's landmark book was published by George Allen and Unwin, Ltd., London, in 1969.) Fortunately, much of the mathematics we need to deal with the functioning of consciousness in a reality both local and nonlocal, was developed by Brown and presented in Laws of Form. His trail-blazing work is acknowledged here with deep gratitude. Its importance in the development of transcendental physics can hardly be over emphasized.

It is interesting to note that Brown was probably working out the details of his Laws of Form at about the same time that Bell was completing his theorem. But it wasn't until the Aspect experiment established the validity of the Copenhagen interpretation, about twenty years later, that it became possible to see the full significance of Brown's Laws of Form.

The mathematical logic of the Laws of Form is as self evident as that of conventional algebra and the calculus of Newton and Leibnitz, and derivation of the Laws of Form is similar to that of the Calculus of Distinctions as presented in Appendix D and can be pursued by anyone with sufficient interest and perseverance. But even without tracing every logical step of the Calculus of distinctions or the Laws of Form, we may understand how appropriate they are as the basis for the mathematical language of transcendental physics.

By recognizing the role of consciousness in the creation and propagation of the objective forms of the universe, we have reversed the basic assumption of scientific materialism. Evidence more direct and less complicated than the logic of Bell and Aspect attest to the validity of this reversal. Looking around us, it is readily apparent that wherever organic life and consciousness is functioning, there exists a higher level of order and structure. Consciousness and matter may be said to be two sides of the same coin. The persistence of increasing order in the vicinity of conscious life is consistent with the hypothesis that order and structure originate in consciousness and are projected into the physical aspects of the universe as a result of the functioning of consciousness. From the other side of the coin, it has been demonstrated many times that abstract mathematical theorems, developed entirely on the basis of logic, with little or no reference to physical reality, are found to have useful application in the physical world. These observations, along with the Bell-Aspect empirical proof of the Copenhagen interpretation, provide sufficient evidence to encourage us in our efforts to reveal the innate structure of consciousness through the development of the Calculus of `Distinctions by tracing the act of drawing a distinction in the substance of consciousness, without reference to the physical universe.

If the physical universe, as we see it now, existed as something complete within itself, devoid of the presence or effects of consciousness, it should exhibit the processes of entropy in the extreme. Without an active organizing force, all structured matter and energy should disappear in a flash. In such a universe, where did the abundance of structure and order that we see originate? We know, of course, that consciousness does exist, and we have presented and discussed compelling evidence and logical proof that consciousness, in an original or primary form had to exist prior to the appearance of the physical universe.

We have also proposed the hypothesis that order, or negative entropy, is a product of the functioning of consciousness. In order to understand how structure may originate in consciousness and to see how it is reflected in objective physical reality, we need to understand the Bell theorem concept of nonlocality and apply G. Spencer Brown's Laws of Form and the Calculus of Distinctions to the conscious drawing of a distinction..

What is nonlocal reality? How does it relate to consciousness, and how does the conscious act of drawing distinctions in the process of observation and measurement participate in the formation of the structure of electrons, protons, quarks or superstrings, photons, particles and waves? By examining our normal experiences of conscious awareness we may begin to glimpse the reality of nonlocality and define the process of the drawing of distinctions. Awareness is localized by focusing attention to local sensation or detail. A sensation such as pain may attract attention, or attention may be consciously directed to a specific spot or area. It is, however, also possible to be aware in a more global sense, without the attention focused on any particular location. Such awareness is nonlocal to the extent that one is simultaneously aware of the existence of all parts of one's body and general surroundings. If we think of this sort of awareness expanded to the scale of the universe, we may be able to envision the possibility of the existence of primary consciousness with universal nonlocality. Given that primary consciousness exists, its nonlocality implies that individual consciousness is connected with it at some level, and thus we would expect individual consciousness to exhibit characteristics at least analogous to those of primary consciousness.

Our purpose here is to explore the structure of consciousness. If our hypothesis is correct, all of the structure in the universe should exist within consciousness. But consciousness functions through the drawing and organization of distinctions. Thus all structures within consciousness are built up of various types and hierarchies of distinctions. Do structures built up of distinctions in consciousness bear resemblance to the structures of physical reality? In the introductory remarks to <u>Laws of Form</u>, George Spencer Brown says:

> By tracing the way we represent ... a severance, we can begin to reconstruct with an accuracy and coverage that appear almost uncanny, the basic forms underlying linguistic, mathematical, physical, and biological science, and can begin to see how the familiar laws of our own experience follow inexorably from the original act of severance. The act is itself already remembered, even if unconsciously, as our first attempt to distinguish different things in a world

> where, in the first place, the boundaries can be drawn anywhere we please. At this stage the universe cannot be distinguished from how we act upon it, and the world may seem like shifting sand beneath our feet.

Can a study of the "act of severance" and the way we represent the act, i.e., the drawing and transformation of distinctions, or, more generally, the Calculus of Distinctions, reveal the universal structures innate to consciousness? While Brown's purpose was not to probe the structure of consciousness, but rather, to reveal the most primary logic in which mathematics is grounded, he did discover that the same laws of form apply to both the pure internal logic of mathematics and the external laws of physics. He observes:

> That mathematics, in common with other art forms, can lead us beyond ordinary existence, and can show us something of the structure in which all creation hangs together, is no new idea.

And further along in the introduction, he says:

> One of the motives prompting ... the present work was the hope of bringing together the investigations of the inner structure of our knowledge of the universe, as expressed in the mathematical sciences, and the investigations of its outer structure, as expressed in the physical

sciences. Here the work of Einstein, Schrödinger, and others seems to have led to the realization of an ultimate boundary of physical knowledge in the form of the media through which we perceive it. It becomes apparent that if certain facts about our common experience of perception, or what we might call the inside world, can be revealed by an extended study of what we call, in contrast, the outside world, then an equally extended study of this inside world will reveal, in turn, the facts first met with in the world outside: for what we approach, in either case, from one side or the other, is the common boundary between them.

I do not pretend to have carried these revelations very far, or that others ... could not carry them further. I hope they will. My conscious intention in writing this essay was the elucidation of an indicative calculus, and its latent potential, becoming manifest only when the realization of this intention was already well advanced, took me by surprise.

I break off the account where, as we enter the third dimension of representation with equations of degree higher than unity, the connexion with the basic ideas of the physical world begins to come more strongly into view. I had intended, before I began writing, to leave it here, since the latent forms that emerge at this, the fourth

> departure from the primary form (or the fifth departure, if we count from the void) are so many and so varied that I could not hope to present them all, even cursorily, in one book.

In these statements Brown reveals the focus and intent of his work. Since he wrote them, the "boundary of physical knowledge" he mentions has been breached by Bell's theorem and Aspect's empirical results. Because of this breach, our focus is now different, in that it includes consciousness as both initiator and medium of the original severance. It is solely because of this difference in focus that the Calculus of Distinctions presented in Appendix D is somewhat different than Brown's development of the Calculus of Indications in Laws of Form. The differences will be apparent to anyone who pursues both derivations. At this point in our discussion, however, the only difference the reader needs be aware of is that our symbol, ┐, is used to describe the act of the drawing of a distinction, rather than the distinction itself.

Regarding the symbol in this way leads to the realization that the act of the drawing of a distinction requires the existence of a conscious agent to perform the act. While it is clear that the symbol we use, when it is placed on a sheet of paper, is a physical object, created for use in the representation of a calculus, leading to complex expressions that may be manipulated by mathematical procedures to portray forms and mutations of forms, it should be equally clear that all distinctions, whether perceived as existing in the brain, as patterns of electrical impulses, as symbols on paper, or as objects in the physical world, must also somehow exist in the consciousness of the agent performing the act of the drawing of a distinction. It is the form and structure of these distinctions, existing in consciousness, that we seek to reveal through the derivation of the Calculus of Distinctions on the pure grounds of logic, without reference to forms perceived as already existing in the physical universe.

It can be argued that the logical structure that naturally flows from the formation of the first distinction, is the structure of consciousness. Using the Calculus of Distinctions to pursue the laws of form in their purest venue, abstracted from all physical concepts, we begin to see that what we are tracing is a map of the logical consequences of the drawing of a distinction. It turns out that this map is nothing more or less than a map of the innate structure of consciousness itself.

One of the ways consciousness projects structure into the physical universe is through the interaction of mind, body, and matter in the form of human beings and other sentient beings. When a conscious entity purposefully manipulates physical objects, (e.g., by building something, writing a letter, etc.) corresponding distinctions have been and are being drawn in individualized consciousness. Whether the objects being manipulated are ordinary objects, experimental apparatus, such as light sources, slits, and shutters, or photons and light waves, the actions are first conceived in consciousness through the process of the drawing of distinctions. This transformation of form and structure from consciousness to matter we can understand. But how does consciousness participate in the formation of elementary particles and waves, electrons and protons, photons, etc.? How does an observation or measurement cause elementary particles and waves to come out of the spectrum of possible states?

Since we are normally only aware of the effects or consequences of the existence of elementary particles, not the particles themselves, and since we certainly are not normally aware of the processes by which they are formed and perpetuated, the form of consciousness that is responsible is clearly not the ordinary everyday consciousness that we enjoy as individuals. We have already established that a primary or original form of consciousness existed prior to the formation of the first particle. We shall see, through application of the Calculus of Distinctions that the innate structure of primary consciousness does include the forms that we observe in physical reality.

An observation, performed by individual consciousness completes a self-referential loop or feedback that confirms a specific pattern that was already inherent within primary consciousness. The appearance of the totally separate existence of the object of the observation is an illusion created in the limited space-time continuum selected from the spectrum of reality by the sense organs of the observer.

A universe devoid of distinctions is both infinite and void, since there are no boundaries anywhere. In a universe constructed of a finite number of distinctions, whether that number is one or billions upon billions, everything that exists, or can exist, is part of one reality. That one reality is infinitely continuous, since there is no limit to the number of distinctions that can be drawn, and the measures of extent in both time and space are directly related to the number and size of the distinctions that are drawn. These observations may be boiled down to two primary statements from which the Calculus of Distinctions can be derived: 1) There is only one reality, and 2) Reality is infinitely continuous. These two assumptions can be rendered symbolically as two equations:

1) ⌉⌉ = ⌉ and 2) ⌉̅⌉ = .

Where ⌉ represents the conscious act of the drawing of a distinction, without regard to the size, shape, or type of distinction, and the blank space on the right-hand side of equation 2 represents the void. For clarity during calculations or derivations, the symbol ⌉̸, implying no distinctions is used to represent the void state, instead of the blank.

Equations 1 and 2 represent the local and nonlocal aspects of reality. This may be seen by reasoning as follows:

Equation 1 may be thought of as representing the essential unity or nonlocality underlying all reality, since the number of distinctions on the left side of the equation is immaterial.

Demonstration:
The statement $\urcorner\urcorner\urcorner \cdots \urcorner\urcorner\urcorner = \urcorner$ is true, no matter how many distinctions are on the left side of the equation, since any pair of distinctions may be condensed by use of equation 1, to a single distinction, and this process may be continued until, in the case of an even number of distinctions, only two distinctions are left and the expression has been reduced to equation 1, or until, in the case of an odd number of distinctions, only one distinction is left and the expression is reduced to $\urcorner = \urcorner$, which is an identity, and therefore true by definition. This application of equation 1 to itself reflects the recurrent theme of self-referentiality that naturally arises as consciousness turns to look at itself. Since equation 1 expresses the essential unity of all distinctions, it also expresses the nonlocality of Bell's theorem. If each distinction in equation 1 is taken to represent a distinction of self from other, i.e., individual occurrences of consciousness, then equation 1 represents the nonlocality described by Erwin Schrödinger when he observed that the "I" experienced by each individual is one and the same.

Equation 2 completes the description of the two basic features of reality by expressing the essence of relativity and locality in non-numerical terms. The act of drawing a distinction involves the creation of a boundary which divides reality into two parts: the part distinguished, and everything else. Relative to that which is distinguished, everything else is part of the infinite void, since no distinctions are consciously drawn in it. This equation may be understood as the symbolic representation of the creation and dissolution of a distinction, or the crossing and re-crossing of the boundary separating the distinction in consciousness from the void. Consciousness occupies the distinction by crossing the boundary from the infinite void into the region defined by the distinction. If the boundary is crossed a second time, the focus of consciousness has returned to the void. If reality is divided by numerous distinctions, consciousness experiences locality by occupying a finite set of distinctions as if they comprise a single distinction. Reflecting this, a local distinction might be represented by the Calculus of Distinctions expression, $E = \overline{\ \ }|$, but E may be comprised of further distinctions such as: $E = \overline{x}|$, $x = \overline{\ \ }|\overline{\ \ }|$, etc. At this point, we can see how complex expressions may be built up by repeated applications and combinations of equations 1 and 2. These equations are the primary expressions of the Calculus of Distinctions from which all the structures and forms of primary consciousness may be derived, as demonstrated in Appendix D.

Returning for a moment to the basic concept of distinction which underlies the Calculus of Distinctions, we note that classifying distinctions as conceptual, perceptual and/or existential is both useful from the theoretical standpoint of understanding the nonlocal, self-referential, and participatory nature of reality, and from the practical standpoint, since it provides the basis for what may turn out to be an important use of the Calculus of Distinctions: the means to determine whether a given expression represents an existential, perceptual, or conceptual form.

It is of particular interest and significance to note the way in which the Calculus of Distinctions is used to describe the structure of primary consciousness. First, the forms represented by the calculus are built up from the primary conscious act of drawing a distinction in an infinite void. Thus the logical structures of the calculus are based on the primary function of consciousness, the drawing of a distinction, which is reflected in acts repeated countless times daily by sentient beings everywhere. Second, the forms of the calculus become more and more similar to the forms of objective physical reality as the degree of self-referentiality or subversion increases. This point was mentioned by Brown in the passages quoted above and was further articulated in Chapter 11 of Laws of Form and the notes related to that chapter as follows:

> Equations of expressions with no re-entry [i.e., with no instances of self-referentiality]... will be called equations of the first degree, those equations with one re-entry will be called of the second degree, and so on.

> Such an expression [second degree] is thus informed in the sense of having its own form within it, and at the same time in the sense of remembering what has happened to it in the past.

The introduction of time allows the Calculus to include self-referential expressions of second or higher degree and equations with two or more solutions, revealing, or remembering the expression's information, i.e., the way in which the expressions were formed. Brown continues:

> We may perhaps look upon such a memory ... as a precursor of the more complicated forms of memory and information in man and the higher animals. We can also regard other manifestations of the classical forms of physical or biological sciences in the same spirit.
>
> Thus we do not imagine the wave train emitted by an excited finite echelon [pages 63 and 64, Laws of Form] to be exactly like the wave train emitted from an excited physical particle ... (We should need, I guess, to make at least one more departure from the form before arriving at a conception of energy along these lines.) What we see in the forms of expression at this stage, although recognizable, might be considered as simplified precursors of what we take, in physical science, to be the real thing. ... For example, if, instead of considering the wave train emitted by the expression in figure 4 (page 64), we consider the expression itself, in its quiescent state, we see that it is composed of standing waves. If, therefore, we shoot such an expression through its own representative space,

it will, upon passing a given point, be observable at that point as a simple oscillation with a frequency proportional to the velocity of its passage. We have already arrived, even at this stage, at a remarkable and striking precursor of the wave properties of material particles.

It is also of particular significance and interest to mathematicians and logicians to note the analogy of higher-order equations in the non-numerical Calculus of Distinctions to the multi-dimensionality of Minkowski or Hilbert space, and imaginary numbers in ordinary numerical algebra. But it is more central to the purposes of this discussion to return to our consideration of the concept of existence. We have noted that for a distinction to be existential, as opposed to purely conceptual or only perceptual or illusory, it must exist in consciousness and extend in at least three dimensions. Now we see that the logical structure of Brown's Calculus of Indications, which is identical with the pure structure of consciousness revealed by the Calculus of Distinctions, begins to mimic the forms of physical reality as soon as equations of any degree higher than unity (the third departure from the void) are introduced.

In this chapter, we have proceeded on the premise that consciousness is virtually the only thing we experience directly. Since the observation of, and knowledge of what we believe to be objective reality is achieved and acquired only through the functioning of consciousness, the development of any real scientific understanding of the nature of reality must begin with a scientific understanding of the nature and functioning of consciousness. In the historical development of science, this basic step has been pretty much overlooked and neglected. In fact, the assumption was deliberately taken, for some quite understandable reasons, but without any scientific basis, that physical reality exists independent of and apart from consciousness. The scientific materialism based on this assumption has enjoyed great success as a tool for discovering relationships between limited aspects of physical reality, but in the end, it produced paradox and conflict, precisely because it ignored, and even denied the existence of the causative and organizing agent consciousness.

It is now time to turn back, investigate consciousness in an objective manner, and develop a scientific understanding of consciousness comparable to our current understanding of physical reality. In order to do this, we must divest ourselves of the notion that objectivity relates only to things that we regard as physical. We have seen in this chapter that the objective investigation of the basic functions of consciousness, the drawing of distinctions and the creation of structure, leads to the realization that consciousness and physical reality are two sides of the same coin. They stand out as the two complementary aspects of one all-encompassing reality.

Western science may be forgiven the error of neglecting the study of consciousness, for this study is intrinsically more difficult than the study of specific, limited physical features abstracted from reality. It is difficult to develop a scientifically objective method for studying consciousness, since to do so requires the application of the observational and reasoning powers of consciousness to the study of itself. The sensitivity and self-involvement of the egos of individualized consciousness are such that it is no wonder that the scientists developing their budding discipline in the Middle Ages chose to limit their focus to the more tractable external physical world. However, science should have matured by now to the point where the paradox and conflict brought about by this oversight can be recognized for what it is and rectified. It has been more than a decade since the resolution of the EPR controversy gave us a clear indication of the integrated nature of conscious reality and physical reality. As scientists, we can no longer allow ourselves the intellectual lassitude of confusing our limited materialistic models with reality.

It is natural that after all these years of focusing exclusively on the material aspects of reality, we might fear that turning inward will lead us to confuse irrational, dream-based pseudo-realities, wishful thinking, and imagination with objective reality. This is an acknowledged danger if an inner study is attempted without scientific and personal discipline. However, when consciousness is understood to be a real substance, as much a part of reality as matter and energy, it is possible to avoid this danger by concentrating on the truly objective aspects of consciousness and thereby achieving a scientific understanding of the nature of consciousness and its relationship to the rest of reality.

Bringing the distinctive powers of consciousness to bear on consciousness itself has been likened to the eye attempting to see itself. But, of course, the eye <u>has</u> learned how to see itself. Natural reflective surfaces, like pools of water suggested ways to do it, and we have developed and improved mirrors to the point where it is possible to see, magnify, and photograph even the interior of our own eyes. Are there mirrors of consciousness? Can we develop and improve them to the point that we can objectively map the details of our own consciousness? The answer to both of these questions is unquestionably in the affirmative.

The Calculus of Distinctions has revealed the structure of primary consciousness, and it is clear that this structure is reflected in individual consciousness, which is actively involved in the projection and perpetuation of the structure of primary consciousness into the physical universe. It is also clear that individual consciousness holds individual memories, thought structures, and dreams, often distorted by various emotions, such as anger, fear, hate, greed, covetousness, and desire, all created by false but often very strong ego attachment to the body and other aspects of the physical world. It is, therefore necessary, in the interest of objective science, to proceed very carefully.

The ancient Greeks peered into dark pools at places like Delphi and saw projections of images from the depths of their own consciousness. Modern psychologists and psychiatrists use hypnosis and other methods of psychoanalysis to probe the inner consciousness of their patients. But such projections usually appear, like dreams, to be complex symbolic representations of layers of mental stress and conflict, and tend to be very subjective. This is not the type of structure we are looking for in an objective understanding of consciousness. This problem of irrational subjectivity arises because individual consciousness reflects the nonlocality and infinite continuity of primary consciousness as expressed in assumptions 1 and 2, and is therefore capable of drawing a virtually unlimited variety of complex distinctions within itself.

It may be difficult, within individual consciousness, to distinguish between distinction complexes containing delusional concepts and those reflecting reality. However important it may be to an individual to recognize, deal with, and dissolve such complexes, in order to initiate an objective study of individual consciousness in the spirit of transcendental physics, we must turn away from such complex structures in the initial development of transcendental physics and focus on our most basic conscious process: the drawing of a single distinction. By doing this, we may begin to reverse the development of the complex multi-distinctional structures that reflect a sometimes distorted image of reality, and return to the original distinction of self from other, in a way very much analogous to the tracing of the expanding universe back to the big bang. This process of achieving objectivity and clarity may carried out on both abstract and concrete levels.

When we do this within our own consciousness, we behold what the Zen contemplatives call our original face, or the primordial Mind. It usually manifests as an objective light perceived within, and is accompanied by profound peace and mental clarity. When this is successfully accomplished, the distorted images created by false ego attachment will fall away, and as we return to our day-to-day existence, we will be able to see pure reality, as it really is, undistorted by ego-based psychoses. When we do this symbolically, and progress forward from the first act of drawing a distinction in a mathematically logical manner, as we did with the Calculus of Distinctions, we are able to develop a language capable of describing the innate structure of primary consciousness. And this, as it turns out, since the structure that we perceive in physical reality originates in primary consciousness, provides us with a map of the true nature of reality.

When we focus on a specific aspect of reality in this manner, the benefits might be very rewarding. For example, the physical body you inhabit as individualized consciousness is a pattern drawn from primary consciousness. This pattern may have become impacted and distorted through stresses encountered inthe physical world, resulting in a weakened physical body. A pattern of increasing sickness or disease might be reversed by superimposing an image of the original pattern of health and strength, in the form of energy fields, on the existing body.

What we seek in transcendental physics is to investigate both the physical and conscious aspects of the world that we experience objectively, in order to arrive at a deeper and more comprehensive understanding of the true nature of reality. To do this effectively, the tools used by physicists must be augmented with improved mathematical tools and with the tools of refined individual consciousness. To this point in our pursuit of transcendental physics, we have determined that structure, form, and order originate in primary consciousness, that individual consciousness participates in the projection of that order into the physical world by interacting with the physical world in various ways, including the act of making observations. And we have developed the mathematical language of the Calculus of Distinctions to deal with the nonlocal, consciousness-based aspects of reality. We will continue, in the next chapter, to explore the nature of reality existing in the eternal structure of primary consciousness, selectively revealed through individualized consciousness, and continuously reflected in the ever-changing physical universe.

"*Thus shall ye think of all this fleeting world: A star at dawn, a bubble in a stream; a flash of lightning in a summer cloud, a flickering lamp, a phantom, and a dream.*"

- Gautama Buddha, India, ca. 400 B.C.

CHAPTER 8. THE DOORWAY OF LIGHT

"With all your science can you tell how it is, and whence it is, that light comes into the soul?"

- Henry David Thoreau

The Experience of Light

The year was 1960. This September evening I made my way alone up the winding way of Mount Washington Drive. A serpentine ribbon of pavement, the road to the top of Mount Washington was lined on the outsides of curves by white wooden guard rails, punctuated with occasional agave marginata and century plants, emblazoned near the top with vibrant purples of moss rose and ice plant. I knew every turn by heart now, having visited the beautiful, peaceful grounds at the top of the drive often during the past year. My college roommate had become a renunciant. He had been in the order for more than a year. The reason for my trip to Mount Washington this evening was to receive a spiritual initiation.

As I stepped through the arched gateway, a tangible peace descended upon me. I walked slowly along the drive, enjoying an aromatic mixture of scents, including bay leaf, gardenia, and sandalwood, wafting on the still evening air. I looked out over the city. A carpet of lights sparkled below, as far as the eye could see. A few minutes later, I joined a small group of expectant initiants waiting in the lobby of the main building. Soon we were absorbed in the details of the initiation ceremony.

Toward the end of the ceremony, we received instructions in concentration techniques. Immediately following the instructions, I beheld an inner light like nothing I had ever seen before. There was nothing vague or dream-like about this light. It was just as real as any light I have ever seen from any external source. It was golden in color, spinning rapidly, and vividly three dimensional. I have learned since that this type of experience comes to most seekers at some point, with the proper preparation, effort, and concentration. The reason I am describing this event in this book is because it is an example of inner objectivity, a clear observation of an objective aspect of consciousness that I have experienced myself.

Awareness and Inner Objectivity

In Choosing Reality, A Contemplative View of Physics and the Mind, New Science Library, Shambhala, Boston & Shaftesbury, 1989, Page 170, discussing the Buddhist approach to refining human consciousness, B. Alan Wallace says:

> As one continues to apply oneself to the practice, (referring to the practice of sitting alert, with spine straight, watching the breath) there eventually arises an acquired sign. To some, this sign appears to the mind's eye like a star. ... This sign may arise when one may remain focused on the breath for roughly an hour with only a few brief conceptual distractions. It arises spontaneously and is NOT

> to be intentionally visualized. When the sign appears regularly and with continuity, one directs the attention away from the tactile sensations of the breath and now focuses entirely on this purely mental image. With this as the object, mental stability and clarity are developed...

Wallace goes on to describe further stages of the refinement of individual consciousness and the development of desirable mental states and powers. It seems to me very likely that the stability and clarity of mind developed, even at this early stage, would enhance the abilities of a scientist to think clearly and logically. As aspiring physicists, we attend institutions of higher learning for a period of from four to ten or more years to learn the concepts and procedures of mathematics and physics, and it may take many years to design, build and refine the laboratory equipment we need to reveal a new aspect of physical reality (as it did with the Aspect experiment). If we are willing to expend this kind of effort to become a physicist, why not look into ways to refine our most important tool and most personal piece of equipment, the mind?

Objective Investigation of Inner Realities

It should be possible to refine the capabilities of individual consciousness to the point that they can be focused on the receptorium to investigate it in an objective manner. Information gained in this way, combined with the studies of the neurologist and the physiologist, should give us greater insight into the functioning of consciousness. There are numerous references in scriptures, mystical writings, and psychic research of the objective appearance of inner light. Paramahansa Yogananda, in The Autobiography of a Yogi, on page 180, while commenting on the biblical passage:

"If therefore thine eye be single, thy whole body shall be full of light" --Matthew 6:22, tells us that:

> During deep meditation, the single or spiritual eye becomes visible within the central part of the forehead. This omniscient eye is variously referred to in scriptures as the third eye, star of the East, inner eye, dove descending from heaven, eye of Shiva, eye of intuition, and so on.

The separation of consciousness from the physical aspects of reality seemed reasonable to early scientists, but it led to the alienation of mankind in modern society from the very roots of existence. Knowing now, as we do, that matter and mind are complementary aspects of the one substance of reality, we can understand why science has been unable to produce a consistent "theory of everything": Everything was simply not included in either the assumptions or the methods of investigation. We have been neglecting more than half of the picture. With the re-discovery of inner objectivity, and the knowledge that the acuity of inner observation can be enhanced by certain practical methods and techniques, we have a powerful new tool, the existence of which I can personally vouch for, that can be used in transcendental physics to investigate the *receptorium* or *nexus* that exists between conscious reality and physical reality.

There have been tens of thousands of books, articles, and papers written by physicists and students of physics. Some are profound, some contain errors, some are mostly error. They are carefully stored in the stacks of university libraries, on film, video tapes, and computer discs around the world. You have to pick and choose from among them to learn what you need to know to become a successful physicist. There also exists a wealth of literature on the refinement of inner objectivity, written by some of the leading scientists of the soul and their students. Western science has largely ignored this literature, almost entirely without investigating its claims, on the basis that it cannot be objective. But when we shake off our materialistic blinders and study these writings with the same diligence that we devote to the literature of physics, testing the proscribed methods for practical results and repeatability, we find that some of the information is profound, some contains errors, and some is mostly erroneous. How is this different from the literature of physics, or any other science? Combining the relevant material from both bodies of literature, the transcendental physicist has a vast repository of information to draw on for clues to the location, nature, and accessibility of the mind-body nexus, the *receptorium*. Let's look at some of the clues.

Physical Science Reveals Clues for Transcendental Science

In Chapter 3 we considered light, the medium of observation, from the physical side and found that an understanding of the nature of electromagnetic radiation is crucial to understanding the nature of reality. The evidence is very compelling that light is the key piece of the puzzle. The constant light speed in the physical universe governs the scale of measurement for time and space relative to the conscious observer. The components of speed are distance travelled (space) per unit of time. Thus, the size of the most basic units of space (Planck's length) and the most basic units of mass, energy, and momentum (functions of Planck's constant and the speed of light) are all dependent upon the observable speed of light.

Light interacts with matter in the form of photons, and the effects of photons give rise to another set of important elementary quanta known as electrons. Electrons play an important role in the formation of matter, the reception and transmission of the energy of light, and the observation of physical reality. The close relationship between the photon and the electron is revealed in Einstein's explanation of the photoelectric effect: Electrons occupy spherical shells that account for most of the volumetric space of atoms where they interact with photons. The energy of one photon is absorbed by a single electron, and when an electron drops to a lower energy shell, a photon of light is produced. This exchange of energy allows information to be transmitted from observed phenomena to the observer at near light speed.

The information needed to construct an image of the world, selected from the spectrum of possible worlds by the limitations of our apparatus of observation, is delivered to the optic nerve by photons. Stimulated by the energy of the photons, the cells of the rods and cones in the retina transmit electron pulses through the optic nerve to the brain, roughly in the following manner: Various specialized cells, called neurons, that exist in the optic nerve, organize information related to intensity (number of photons), color (wave length), and changes in these over time. The structure of each neuron allows it to select quanta of energy in a way that encodes the specific types of information needed to form images. These quanta of energy are transmitted from neuron to neuron, along tiny fibers called axons and dendrites, and organized into meaningful pulses by the synaptic relays between the connecting fibers. The synapses are designed to discharge only when a specific threshold of electron energy (a specific number of quanta) is reached. Thus the selected information is delivered to the brain in a meaningful code of quantized impulses.

The nerve cells in the brain function like a combination of digital and analog computers, dealing with the bits of information in the form of structured pulses of electron charge. In terms of the Calculus of

Distinctions, a charged quantum can be considered as a distinction or a simple set of distinctions. And we know that, at some point, per the Copenhagen interpretation, each individual quantum must register its effect on a receptor. If the receptor is a neuron, it must pass the effect of the quantum along in the form of another potential quantum, but this quantum must also have a receptor to register as a physical effect, and so on, until the non-quantum receptorium is reached. The non-quantum receptor is thus necessary to avoid the impossibility of an infinite number of physical receptors that would swell the brain to an infinite size, since the quanta of any physical receptor are of a specific finite size, dictated by Planck's constant and the speed of light.

Science at the Threshold

And so, physics has brought us to the doorway. The universe, up to this point, has consisted of complex patterns of quantum effects, selected out of the great smokey dragon of the universal probability wave function by a finite but complex pattern of semi-stable matter/energy waves, making up the sense organs and brain, which particles were themselves, also selected out of the universal probability wave function of primary consciousness. But the mental images that appear as a result of this process are made of a more subtle light. In fact, everything on the other side of the doorway exhibits a profound difference from the matter and energy world of quanta that we know as the physical world. This profound difference is revealed in the fact that every form and structure appearing in the entire physical universe from the doorway outward, is made up of the finite effects of particles and waves, while the reality on the other side, cannot be composed of quanta, and is nonlocal and infinitely continuous. On this side of the doorway, all the images of people, places, and things seen in an entire lifetime can be stored in a space the size of the space occupied by an elementary particle on the other side, and still be recalled in detail.

Perception: Images in Consciousness

The images inside the doorway are of two types: constructed and actual. The constructed images are forms that are consciously constructed from the information as it is received from the neural network of the senses, and/or from storage within the brain. The actual forms are structures intrinsic to primary consciousness. When patterns encoded in quanta of energy, filtered through the sense organs and the neural network, arrive at the receptorium, the actual forms and structures, existing in deeper layers of individual consciousness, act as templates for the construction of an image from the in-coming pulses. This is the function of what we have called the gestaltenraum or image-formation space. If the information coming from outside is incomplete or slightly garbled, stored information can be used to fill in the gaps to complete the picture. This is why we don't see the blind spot existing in our field of vision due to the juncture point of the optic nerve at the back of the eye, and why we sometimes see something different than another observer might see when looking at the same scene.

Most of the time, the receptorium is so flooded with in-coming information from all the senses, that the attention is constantly directed toward the receptorium and the mental energy is absorbed in the activity of the gestaltenraum, so that there is normally no awareness of the actual forms, which link individual consciousness with primary consciousness. This is why concentration techniques such as those practiced in Zazen, Yoga meditation, or Christian mystical contemplation are used to turn the attention away from the sensory flood from the outer world, and focus it toward the contact with actual images in primary consciousness.

From the phenomenological point of view, a conscious observation is the act of selecting an aspect or aspects of reality from the spectrum of available information flooding the universe, in the form of a probabilistic wave function, for interpretation and decision making. But from the point of view of ontology, i.e., primary consciousness, a conscious observation is the completion of a loop in the substance of consciousness extending from primary consciousness, through projected phenomena and individual consciousness, back to primary consciousness itself.

Attachment to External Forms Yields Suffering and the Fear of Death

As conscious, individual beings, we are constantly in a state of mental stress because of the conflict between the natural desire to participate in the temporal play of phenomena, where we have the illusion of freedom and control, and the desire to merge with the timeless nonlocality of the primary consciousness. The attachment or psychological dependence that we have developed for the physical body and attendant physical sensations, such as those related to food, sex, and power, lead us to believe that the finite, ever-changing world, created by the acts of the drawing of distinctions, is the only reality. Because of this belief, we have a great fear of merger with primary consciousness, which is the only thing that can ultimately alleviate the stress and bring lasting happiness. The epitome of this fear, based on ego attachment, is fear of the loss of personal identity, i.e., the fear of ultimate death as the extinction of consciousness in the destruction of the physical body. Limiting one's awareness exclusively to the identification with a physical body is the root of all suffering and leads to an inordinate and unreasonable fear of death.

The Conservation of Consciousness

Fortunately, this fear is dispelled by the realization that consciousness is not dependent upon the quanta of the physical world. In fact, as we have seen, there is empirical evidence and mathematical proof that the quanta of the physical world are actually dependent upon the functioning of consciousness. Since primary consciousness is universally nonlocal, memories stored in brain cells are also stored in consciousness. Empirical evidence from inner and outer science reveals reality as a spectrum of substance grounded in primary consciousness. Matter, at the most localized end of the spectrum, is the most ephemeral and subject to change. As you progress through more and more nonlocal forms, toward energy and consciousness, the substance of reality becomes progressively stable and changeless. Since primary consciousness is universally nonlocal, it follows that every observation, thought, or image ever formed in *any* conscious individual's *gestaltenraum* exists forever in primary consciousness. Thus the fear of absolute death is unfounded.

Even before we had empirical evidence and mathematical proof, the truth of the indestructibility of consciousness was being proclaimed by the real scientists of inner research from their direct knowledge and experience. Jesus said:

> Verily, verily, I say unto you, He that heareth my word, and believeth on him that sent me, hath everlasting life, and shall not come into condemnation; but is passed from death unto life.
>
> Verily, verily, I say unto you, The hour is coming, and now is, when the dead shall hear the voice of the Son of God: and they that hear shall live.
>
> For as the Father hath life in himself; so hath he given to the Son to have life in himself;

And hath given him the authority to execute judgement also, because he is the Son of man.

Marvel not at this: for the hour is coming in which all that are in the graves shall hear His voice.

- From The Gospel According to St. John, Chapter 5, verses 24 - 28.

Regarding the essential nature of human consciousness, Paramahansa Yogananda said:

Man's essential nature is formless omnipresent Spirit. Compulsory or karmic embodiment is the result of *avidya*, ignorance. The Hindu scriptures teach that birth and death are manifestations of *maya*, cosmic delusion. Birth and death have meaning only in the world of relativity.

- The Autobiography of a Yogi, Paramahansa Yogananda, SRF, Los Angeles, 1974.

And Ramana Maharishi, a sage born in Southern India in 1879, said:

The body dies but the spirit that transcends it cannot be touched by death.
The scriptures declare: "So long as the individual regards the corpse of his body as 'I'

> he is impure and subject to various ills such as birth, death, and sickness."
>
> ...The Being-Consciousness-Bliss of the Self does not perish with the body, breath, intellect, and sense organs any more than a tree does with its leaves, flowers, and fruit.

- From The collected Works of Ramana Maharshi, edited by Arthur Osborne, Samuel Weiser, Inc., New York, 1972.

Mystics of all ages and all parts of the world have expressed the same conviction. Edgar Cayce, the American "sleeping Saint" of the first half of this century said:

> The physical body is not the consciousness. The consciousness of the physical body is a separate thing. There is the mental body, the physical body, the spiritual body. As so often has been given, what is the builder? MIND! Can you burn or cremate a mind?

- From a reading by Edgar Cayce on page 123 of Venture Inward, Hugh Lynn Cayce, Paperback Library, Inc., New York, 1966.

Archetypal Forms in the Structure of Consciousness

During the summer of 1964, while I was a graduate student in theoretical physics at the University of

Missouri at Rolla, I took a job with a seismological crew doing geophysical prospecting in the Texas Panhandle. We were headquartered in the small town of Canadian, Texas. Each morning we got up at daybreak and drove out to the areas we were surveying. One such morning, riding in a pickup truck with the Chief Surveyor, I was augmenting the previous night's sleep by dozing as we cruised across the flat, almost treeless Panhandle landscape. We had been driving for some time when, as the truck slowed for some reason, I began to awaken and became semi-conscious of my physical surroundings. I was aware of being in the truck, but my mind seemed to be suspended somewhere between a state of sleep and wakefulness. One thing caught my attention: there was no sound. The input from the audio neural network was not being received by the *receptorium*. It was as if the sounds of the world had been switched off and I was gliding along with the truck, in complete silence, just above the rippling grasses of the prairie.

I opened my eyes, and what I saw sent an electric shock through me. Standing beside the road, about 100 yards ahead of us, was a huge, shaggy buffalo. Buffaloes were not too common in those days, in fact they seemed to be edging toward extinction. Only a handful still survived in small herds in Montana and Wyoming. I had only seen a live buffalo once, a sad, mangy looking specimen in a roadside zoo. But what sent the shock through me was not the presence of a lone buffalo, but the fact that this animal was about 45 or 50 feet tall! I could see the flaring nostrils and wind-reddened eyes. He shook his grizzled head and pawed the earth, as if about to charge at us. I jumped, and suddenly, the roaring sounds of the tires on the blacktop, the wind in the wing window, and the truck radio burst in upon me. The buffalo reared and was transformed into a green leafy tree, doubtless planted there by the state Highway Department to break the monotony of the wind-swept terrain. I glanced over at the driver. He was humming along with the country-western tune blaring from the radio and sipping coffee from a plastic mug, unaware of my startling experience.

More than twenty years later, in the spring of 1995, returning from a camping trip on the edge of the Current River Ozark National Scenic Riverway Area in South Central Missouri, we stopped to open a gate on the Golden Trails Ranch and there, standing beside the dusty track, was a handsome dark brown bull buffalo, still shaggy in his winter coat. His flaring nostrils and red eyes brought the memory of that summer day in the Texas Panhandle vividly back to mind. He shook his shaggy mane, pawed once, then licking his nose with a long black tongue, he turned away with a contented sideways glance, to graze on a clump of lush grass. There are lots of buffalo now; not the vast herds of the pre-European settlement days, but nevertheless, they're back in strong numbers throughout some of the same areas that were once their native habitat.

What I had seen that morning in 1964 was not a dream or hallucination, but an archetypal form, imprinted in the primary consciousness of the High Plains, a symbol, appearing in the *gestaltenraum* of my consciousness of the enduring consciousness of the buffalo in that region. Such forms are always there, available to our inner vision when our attention is withdrawn, either by accident or design, beyond the door of the *receptorium*. I have seen other such forms, usually in natural settings, along river bluffs, on seashores, on rocky mountain tops, and in deep forests.

Passing Through the Door of Light

Returning to our exploration of the *receptorium*, if we can look past the neural networks of the brain, into the matter-mind nexus, which occupies no physical volume, and yet opens into the vast realm of nonlocal primary consciousness, wherein the entire universe and all space-time float like a mote in an ocean of light, perhaps we can conceive of some of the order that exists there.

The sense-organ selected, neuron reduced patterns of electron energy arriving at this subtle interface are received and mirrored in the infinitely continuous substance of the *receptorium* by images reflected from primary consciousness in the *gestaltenraum* function of consciousness. The receiving and imaging features of consciousness are not really separate, the way the physical eyes and brain are. It is more accurate to say that the same nonlocal substance, interfacing with specific physical phenomena, performs both functions.

The Nature of Reality

The nature of reality, then, is both discrete *and* continuous. It is superficially discrete in atoms, stones, planets, stars, and living creatures, but ultimately deeply continuous, because it is everywhere undergirded by consciousness. And individual consciousness is the connection between the transient features of the "outer" world and the changeless structure of nonlocal primary consciousness. At the discrete level, on the surface of reality, phenomena are relatively probabilistic, since individualized consciousness and individual particles have a certain measure of freedom. But at a deeper level, reality is absolutely deterministic in the sense that the detailed form and structure of a loop of observation are absolutely determined by the underlying form in primary consciousness that corresponds to the specific observation that is selected by the observer.

As conscious observers, we participate in the flow of logical structure and order, out of primary consciousness into the still-developing physical universe. We participate by drawing distinctions and making observations. Light is the medium of those observations and the link between consciousness and matter. How does the light of the outer world relate to the light of primary consciousness? Physical light from the sun, or any other material source, is part of spiritual light in exactly the same way that individual consciousness is part of primary consciousness. An individual photon is analogous to an individual sentient being. It exists as a discrete particle defining time and space by its motion, while connected by its continuous spectrum of probable states to the infinitely continuous light substance of primary consciousness. Individual consciousness literally creates its own temporal and spatial dimensions, bringing individual photons out of the wave of probable states by creating distinctions in the act of observation.

We've called the *receptorium* the "doorway of light" because of the central role of imaging and visualization in the definition of our universe of experience. But the attendant information from the other senses, also delivered in the form of patterns of electron energy, adds to the reality of the total experience. All the forms of objective reality are created through the joint action of individual and primary consciousness at the *receptorium*, the nexus where all worlds meet. We are the eyes and ears, hands and feet, inhalation and exhalation of primary consciousness. Reality is not the multitude of changing shapes and forms that appear in a time-space continuum, reality exists as an infinitely continuous, undifferentiated substance all around us. It is only the actions of the conscious drawing of distinctions in the *gestaltenraum* that create the illusion of a universe or universes separate from the infinite whole.

Toward a New Paradigm

In the previous chapters, we have seen how a non-quantum, and therefore non-physical receptor is necessary for the existence of an objective physical universe, and we have seen how an objective investigation of this receptor and its functioning leads to a new, more comprehensive understanding of the nature of reality and our involvement in it. In the next chapter, we will see how the integration of consciousness into both relativity and quantum theory resolves the conflict between them and paves the way for the development of a sweeping new scientific paradigm.

CHAPTER 9.
THE CONVERGENCE OF EXTREMES

If the doors of perception were cleansed
Everything would appear to man as it is,
Infinite.
- William Blake

Mathematicians have defined several types of infinities. But physicists, in general, abhor infinities. They are quite embarrassed when infinities show up in quantum field equations describing real phenomena, because, as every good physicist knows, Blake not withstanding, infinities just don't exist in the *real* world. So one might suppose that a mathematical physicist would be schizophrenic: as a mathematician, he has to believe in the reality of infinity, but as a physicist, he can't. Perhaps if he's careful, he can keep these domains separate. But that's almost as hard as being a member of the Religious Right and a Marxist at the same time. The infinities that do show up, as they are wont to do, are simply eliminated by replacing them with reasonable, finite numbers, numbers obtained from experimental measurements or estimated from heuristical considerations. This process is called "renormalization" which sounds a lot better than admitting that you are inserting a "fudge factor" or cheating, to make the model work.

The differences that divide relativity and quantum physics can be traced to the way they deal with certain infinities. For example, relativity accepts the infinitely small, but rejects the infinitely large. A photon and other elementary particles have no dimensions in relativistic equations. They act as mathematical points. They are, therefore, infinitely small. The universe, on the other hand, is large, but finite. Space is relativistically warped by all the mass in the universe and therefore, a closed, finite system. Any straight line, like the trace of a line on a sphere, must return to its starting point. Thus, according to Einstein, if light were infinitely fast, and nothing else were in the way, all you could see would be the back of your head. Quantum physics, on the other hand, rejects the infinitely small, --all quanta are finite in size and cannot be divided -- and accepts certain types of infinitely large quantities. These differences in the conceptualization of infinities lead to functional differences in applications and the differing ways relativity and quantum physics view reality: Relativity sees reality as continuous and deterministic, quantum theory sees it discrete and probabilistic.

One of the fortuitous benefits of transcendental physics is that it should greatly reduce the incidence of schizophrenia in scientists. In our examination of the *receptorium* in the previous chapter we saw that reality is both discrete *and* continuous, both absolutely deterministic *and* relatively probabilistic. Primary consciousness, the substrate of reality, and individualized consciousness are nonlocal and infinitely continuous. This means that any distance can be divided into halves, thirds, fourths, etc., indefinitely --just like the real number line of mathematics. Physical quanta, the surface phenomena of reality, however, are finite and discrete because all distinctions are finite and limited by the size of the quanta of the observing apparatus. Transcendental physics, using the Calculus of Distinctions, integrates and describes both the finite discrete and infinitely continuous aspects of reality. Continuous and discrete phenomena are understood as not mutually exclusive, but complementary aspects of reality. The continuity of consciousness and the discreetness of matter and energy are also seen as complementary interacting aspects of reality.

The basis of the application of the Calculus of Distinctions is the concept that a distinction drawn by a conscious entity is essentially a self-referential act involving the selection and separation of some part of primary consciousness from itself. The knowledge of the essential oneness of all reality, including the observer and the observed, is hidden from the observer by virtue of the fact that the observer has severed conscious connection with primary consciousness by the very process of the drawing of distinctions. This severance creates the parameters of extent which we experience as time and space.

Representing the forms and structures of the physical universe as complex expressions made up of many individual distinctions in the primary substance of consciousness leads to a totally different understanding of time and space. The logical and mathematical steps that lead to this new understanding of time and space are presented in more detail in Appendix D. Only the conceptual aspects will be presented here. As individualized consciousness, we can create distinctions conceptually and actually, or we can recognize already existing distinctions. By defining elementary particles of matter and energy as the ultimate physical distinctions, we are able to apply the logic of G. Spencer Brown's Laws of Form to physics and the Calculus of Distinctions becomes a basic language capable of describing the quantum world as the physical aspect of the broader reality of primary consciousness.

In the Calculus of Distinctions, matter and energy, time and space, in short, all the details we associate with the objective universe, are related to the conscious acts of making observations, i.e., drawing distinctions. We have discovered that the universe had to have been formed by a primary act of distinction. We are accustomed to dealing with complexes of distinctions which we experience as already existing. But how can we conceive of a reality that existed before the first distinction was drawn? We can't imagine an infinite expanse of nothing, but we can imagine an equation describing a range or spectrum of possible states. Such an equation is the Schrödinger wave equation. Imagine primary consciousness as nonlocal potential, a set of possibilities, any one of which can be actualized by the presence and actions of an observer. We have shown that non-quantum receptors exist within the physical structure of conscious observers, but at the beginning of the universe, where did the first observer come from?

A Czech logician name Kurt Gödel became famous by proving that there are always questions that can't be answered within the structure of any logical system. We've always known this, but it was very important to have it proved. To see why we've always known this, consider the following example: When I was in high school, the highlight of the day for me was always the math class that contained both myself and a friend named Marvin Forbes as students. We loved to debate every aspect and detail of any geometric proof, trig, algebra, or logic problem. We were always trying to stump each other. One day, one of us came across a simple question that illustrates Gödel's theorem: "A man jumped off a bridge. Where was the man when he jumped off the bridge?" This seems like a perfectly logical question to ask. All of the terms are well defined, the syntax is correct, and the question is clear. But there is no answer. Where was the man when he jumped off the bridge? On the bridge? No, that was before he jumped. In the river? No, that was after he jumped. In the air between the bridge and the river? No! That was also after he jumped. Attempts at other answers take us into the realm of the absurd. "He wasn't anywhere!" My friend declared, "We've exhausted all the places he could possibly be!"

The question "Where did the first observer come from?" is not answerable in any system of logic dealing with an ordinary local reality of time, space, matter, and energy. We know that there had to be an observer before any form could be precipitated from the nonlocal continuum of primary consciousness. But a reality of pure primary consciousness with no distinctions requires a different system of logic. There is an answer, of course: the first observer had to be primary consciousness itself. And we know that a non-physical, nonlocal receptor is both possible and necessary to the existence of the universe as we experience it. See Chapter 2 and Appendix C. But the way in which primary consciousness formed the first receptor is necessarily a matter of speculation until we can discover a way to recover the memory of it from primary consciousness. The main problem of beginnings and endings, however, may be simply a function of our concept of time. We shall see that a more comprehensive understanding of time will remove the paradox.

Wheeler suggested that the choice an observer makes in determining the path of a photon is a "primordial act of creation". This means that time, in totality, must be something different, something more than the time we experience. The Calculus of Distinctions provides the solution. Recall from Chapter 6 and/or Appendix D that to exist in reality, a distinction must exist in consciousness and in at least three dimensions. An extension of one dimension is an abstraction. In reality, something with only one- or two- dimensions turns out to be merely a conceptual aspect of a real three-dimensional object. Such an object can only actually exist as part of a three-dimensional feature of reality. We saw that time enters into expressions of the Calculus of Distinctions as a fourth dimension of extension. But time itself, as we experience it, appears to be uni-directional and linear, i.e., one dimensional. The mathematics of the Calculus of Distinctions suggests that in order to exist in primary consciousness, time must be three dimensional. But how do we visualize three-dimensional time? Actually, based on what we've already learned about primary consciousness and its inherent nonlocality, it turns out to be quite easy.

Primary consciousness is universally nonlocal, as opposed to the limited nonlocality of individual consciousness. this means that primary consciousness is simultaneously present at all points. Thus, time for primary consciousness is like time for the photon, there is no motion, and consequently, no time. It follows, therefore, that primary consciousness is nontemporal as well as nonlocal. Not only is primary consciousness everywhere, it is also <u>everywhen</u>. But, you may ask, if primary consciousness is all-pervasive, and acted as the first receptor, why does it need individualized consciousness, why wouldn't it cause the precipitation of all elementary particles, atoms, trees, etc. anyway? The answer is that *it does*, but what do you have when every possible state of every possible physical particle is precipitated? You have the undifferentiated state itself. It takes the limited nonlocality of individualized consciousness to separate out the distinct timelines and accompanying space that make up finite, physical universes.

Bell's theorem and the Aspect results require nonlocality, and we can envision limited nonlocality because we experience a limited nonlocality in our own individual consciousness. We see the physical reality of space as three-dimensional, but because our sensing apparatus is physically limited in a way that causes us to select only one of the many possible local manifestations of an object such as a photon, we only experience one dimension of time. In Chapter 3 we concluded that time and space are created by observation, and being universally nonlocal, primary consciousness encompasses all things. Thus, in primary consciousness, time has to be three dimensional to accommodate all of the events in each of the many space-time continuums that are available for selection by all observers.

In order to see this more clearly, recall the example of relative motion in Chapter 3, where two observers are approaching each other and a light source at a large fraction of the speed of light. In order to avoid ambiguity, we had to conclude that both observers could never be observing the same physically local photon. Each one was selecting a different photon from the spectrum of possible photons. We can now see that by this act of observation, each observer was also selecting his own space and time. Each observer's space-time continuum is internally consistent and thus there is no space-time paradox. If, in addition, we see that each observer selects his own timeline out of an infinite number of possible timelines in three-dimensional continuum of time, the whole picture becomes clear.

Three-dimensional time allows primary consciousness to contain not only all the events observed by one individual observer in past, present, and future, but those of all observers. Three-dimensional time contains the time coordinates of all the possible events that can be selected out of the spectrum of the universal wave equation by all possible observations. From the viewpoint of primary consciousness, everything, all possible past, present, and future events are occurring in the ever-lasting present. From this vantage point, all the timelines of all individual conscious observers appear like vectors in three-dimensional time, completing the perfect symmetry and continuity of primary consciousness.

From this perspective, the question of the first observer dissolves. All events, observers, and objects of observation exist in primary consciousness in the present. Everything that ever existed, now exists, or will ever exist, exists RIGHT NOW in the complete reality of primary consciousness. The appearance of beginnings and ends are illusions perpetuated by our reliance on physical sense organs. The nonlocal global awareness of primary consciousness encompasses all time and space, and thus, like the real number line of mathematics, has no beginning or end.

The primary act of creation is the drawing of a distinction. The creation of <u>this</u> as different from <u>that</u> severs nonlocal reality and results in the appearance of opposites and extremes. Any given finite, four-dimensional universe is a partial view of reality, selected through individual observation by a sentient observer, and it sustains the illusion of separateness and the appearance of a world of duality: dark and light, heat and cold, space and matter. But in the reality of nonlocal primary consciousness, the extremes converge.

The Calculus of Distinctions provides for the first time, a consistent mathematical framework for several innovative theories that have been put forward to explain some of the puzzles of quantum mechanics. The recognition that the time lines of all possible observers form a three-dimensional time continuum is especially helpful. We'll have a brief look at two such theories: Hugh Everett's theory of the universal wave function; see <u>The Many-Worlds Interpretation of Quantum Mechanics</u>, B.S. DeWitt and N. Graham, editors, Princeton University Press,

Princeton, 1973; and John Cramer's transactional interpretation of quantum mechanics, published in a series of papers at the University of Washington, Seattle in the 1980s.

Let's look at the "many-worlds" interpretation first. The idea of three-dimensional time has, implicit within it, a reality of many separate worlds or parallel universes. To see that this is the case, recall again the relative motion example with multiple observers and a plurality of photons, from Chapter 3. We have established the fact that the observers of this example are on different timelines and that they see different photons. The photon selected from the continuous wave function by each observer exists on its own timeline in three-dimensions of space. But these dimensions are described by different sets of coordinates, in the separate space-time continuums of the different observers. A little exercise in the application of the Lorentz transformation equations will quickly show that, just like the location of the photons in the Chapter 3 exercise, no unambiguous relationship exists relating the time coordinate of one observer to that of another and back. Thus each observer functions within his own four-dimensional space-time continuum, or in short, in his or her own separate universe. See Appendix E-II for a detailed demonstration of this.

In Parallel Universes, Simon and Schuster, New York, 1988, Fred Alan Wolf states the case for parallel worlds on page 21 very strongly as follows:

> Quantum physics also indicates a new effect -- the effect that an observer has upon a physical system. This effect cannot be objectively understood without the existence of parallel universes.

And parallel universes cannot be properly understood without the concept of three-dimensional time. The delayed-choice experiment described in Chapter 2, was originated by John Wheeler as a thought experiment, but it was actually carried out by Alley, Jakubowicz and Wickes at the University of Maryland in 1985. The results confirmed the fact that in the overall scheme of things, time is something very different than the every-day linear sequence that we seem to experience. With the understanding we have gained of the way in which each observer selects a physical universe from the infinitely continuous reality of primary consciousness, and with the Calculus of Distinctions concept of three-dimensional time, we are now ready to present the complete answer to the question of how individual consciousness interacts with reality in the act of observation.

All forms and structures, including the physical bodies of all sentient beings, exist within the three-dimensional time and multi-dimensional space of primary consciousness. Evidence such as the discovery of virtual particles and anti-particles, and the Casimir effect tells us that we are surrounded by a timeless, infinitely continuous sea of surging energy. But the physical sense organs act as selective filters, reducing the input to individualized consciousness to the point that a finite world of only three spatial dimensions and one time dimension is perceived. Also, the form and structure of all things, including individual sentient beings, originate in primary consciousness. Thus an observation is an energy loop, a feedback. In fact, the mechanism of an observation can be understood and explained in terms of energy feedback in three-dimensional time, and this brings us to John Cramer's transactional interpretation of quantum mechanics, mentioned above.

The key, as suggested in Chapter 3, lies in understanding light, or more generally, the nature of electromagnetic radiation. And this is not the first time that delving into the nature of light has "shed light" on the nature of reality. Einstein's visualization of light waves, inspired by his study of Maxwell's equations, led to his discovery of the theory of relativity. Mathematically, Maxwell's equations, formulated nearly 150 years ago, already contained the essence of the special theory of relativity, because they were invariant with respect to the Lorentz transformations. And although Lorentz probably didn't realize it himself, the Lorentz transformations reflect the fact that the speed of light is the same for all observers, regardless of relative motion. Paul Dirac suggested that Lorentz may have thought of this possibility, and thus could easily have come up with the theory of relativity before Einstein, but was simply not brave enough to put forward such a strange concept. But, as we've seen, this simple fact of constant light speed means that time and space can be, and often are, different for different observers.

It seems that Maxwell's equations were years ahead of their time. These amazing equations contain yet another clue to the nature of reality; specifically, a clue to the nature of time: It happens that there are actually two sets of solutions to Maxwell's equations: One set, the well-known solution, describes waves moving spherically away from an accelerated electrically-charged particle. The second set, just as valid mathematically, describes waves moving *backward* in time, to be absorbed by a particle. This second set of solutions has been ignored as a superfluous mathematical artifact because our every-day experience of fragmented reality leads us to believe that movement in time is a one-way affair. Within the framework of three-dimensional time and primary consciousness, however, this backward-in-time solution becomes meaningful. Recall that Schrödinger's wave equation also describes waves moving at the speed of light. What if they also move both forward and backward in time? This is exactly what John Cramer proposed nearly ten years ago in an article entitled "The Transactional Interpretation of Quantum Mechanics", in Reviews of Modern Physics No. 58 (1986). Does the Schrödinger wave equation have reverse-time solutions?

When Schrödinger first developed the wave equation to describe the motion of electrons around the nucleus of an atom, he was very careful to include the effects of relativity on the motion of the electron. When he checked the resulting predictions against experimental observations, he was shocked to find that they didn't match. When he removed the relativistic considerations, the predictions matched the experimental results reasonably well. After some cogitation, he left the relativistic factors out and published the results. As it turned out, the problem that caused the discrepancies was not the inclusion of relativity at all, but the exclusion of electron spin, a concept not very well developed at the time.

During his speech in acceptance of the J. Robert Oppenheimer Memorial Prize (Published in The Development of Quantum Theory, Gordon and Breach, Scientific Publishers, Inc., New York, 1971), Paul A. M. Dirac related this story about Schrödinger as an example to illustrate the point that great hopes and fears accompany the development of new ideas in science. Speaking about his own discovery of the anti-electron, or positron, and the way science is advanced by individual efforts, Dirac said:

It is when one is challenging the main ideas that one

> has the great excitement and the great fears that something will go wrong and this sort of excitement has not recurred since those early days. (He referred to a few years following 1925 as the 'Golden Age of physics'.) ... we are all hoping that a new Golden Age will appear, triggered off by some very drastic new idea and leading once again to a period of rapid development with great hopes and fears.

Because of his fear of being wrong, Schrödinger left the relativistic considerations out of his famous wave equation. Students of quantum mechanics are introduced to the wave equation in a simplified, one-dimensional, non-relativistic form. They are told that the equation they become familiar with is a simplified version, and that the complete and correct form of Schrödinger's wave equation must include relativistic factors. But this information often gets lost or forgotten. When the whole expression, complete with relativistic effects, is written, Schrödinger's equation resembles Maxwell's equations, and just as with Maxwell's wave equations, two sets of solutions appear, one describing waves moving forward in time, one describing waves moving backward in time! John Cramer's insight was that these "secondary" solutions could help explain some of the puzzles of quantum mechanics.

In Cramer's transactional interpretation of quantum mechanics, each elementary particle, e.g., a photon or an electron, produces waves both forward and backward in time. When the forward wave of a source finds a receptor, the receptor responds with a backward wave. The forward and backward waves cancel each other everywhere in the space-time continuum of the universe in which they are distinct, except along the direct path between the source and the receptor, and the "transaction" is completed. How can waves of probability move backward in time? Cramer's explanation involves viewing the situation "atemporally", i.e., outside of time. But this is simply a word game, a semantical ploy that becomes unnecessary when we realize that time is actually three dimensional and that events not on a given timeline are not atemporal, they're simply on another timeline. Cramer was not willing to propose anything so radical as considering backward time as a reality. In fact, he emphasized that this interpretation does not yield results any different from those produced by orthodox quantum mechanics, and proposed that transactional analysis should be used only as a conceptual model to help us understand quantum phenomena.

The many-worlds and transactional interpretations were each offered, by Everett and Cramer as conceptual aids to understanding "quantum reality". In fact, they are perfect examples of the way we tend to ignore mathematical results that do not correspond to our preconceived notions of reality. The basic mathematics underlying relativity and quantum mechanics, i.e., the field equations of Maxwell, Schrödinger, and Dirac, all indicate that reality is more complex than the world we perceive through the senses, and they all specifically indicate that time is not the uni-directional, one-dimensional parameter we take it to be. But both Everett and Cramer, like Wheeler and Bohr, (See Chapters 2 and 5.) were careful to stick to the party line of the paradigm of scientific materialism and disallowed the involvement of consciousness in the "actualization" of physical phenomena from the spectrum of possible states of the wave equations.

According to these theorists, we can never experience parallel universes, since the universe we inhabit is actualized only by events that occur in this universe. And reasoning that transactions involving backward movements in time are atemporal, Cramer had this to say about observer involvement:

> The issue of when the observer decides which experiment to perform is no longer significant. ... the fact that the detection event involves a measurement is no longer significant, and so the observer has no special role in the process.

With the semantic maneuver of invoking atemporality, Cramer removes the process of quantum transaction from the realm of human activity and from experimental verification. While his conclusion is consistent with the prevailing materialistic paradigm, it only holds up when a single transaction is isolated from the rest of reality. But, if we allow backward movement in time, we can't just stop with one transaction. Everett only identifies the receptor as sending a reverse-time confirmation wave back to the source quantum. The vibrating source in Everett's example must also send a wave backward in time, to its point of origin. There is no way we can avoid the conclusion that the backward waves continue all the way back to the big bang. And so we see, when put into the context of the continuing source-receptor interaction in the real world, the assumption that consciousness has no special role in the transaction fails to survive the logic of infinite descent. While the wave collapse in a given transaction may be caused by the pre-existence of a physical receptor, the elementary particles of that pre-existing receptor had to have had a receptor themselves, and so on, leading to the infinite descent discussed in Chapter 2 and Appendix C. Ultimately, at some point in time, a non-quantum receptor *had to be* involved.

The mathematics of transcendental physics, i.e., the Calculus of Distinctions, unveils a reality of parallel universes on separate timelines, where movement from one timeline to another, or even backward movements are not prohibited. With this understanding, we can see reality in a broader, more complete perspective than that afforded by Everett and Cramer. When consciousness is brought into the equation, as we have done with the Calculus of Distinctions, we find that the existence of a primary form of consciousness is necessary for <u>any</u> physical universe to exist, that time is three-dimensional, and that three-dimensional space, single timeline continuums are limited aspects of reality, precipitated from the complete spectrum of infinitely continuous reality ultimately by observation.

By re-visiting the double-slit, delayed-choice experiments and the EPR paradox, we can see how the Calculus of Distinctions concept of three-dimensional time explains the puzzles and resolves the conflicts of relativity and quantum mechanics. We'll look at the combined double-slit, delayed-choice experiment first.

Referring to Figure 3, in Chapter 2, recall that photons were released one at a time from the source. With the photographic-plate venetian-blind open, each individual photon registered in one or the other of the collectors, indicating the slit through which it appeared to have passed. With the venetian blind closed, the photon impacts formed the characteristic interference pattern, no matter how much time elapsed between the release of individual photons. The puzzling aspects of this experiment, which spawned so much debate, were: How could a decision, made after the photon had passed through the slits, determine whether it behaved as a particle or wave while passing through the slits? And how could a single photon go through both slits and "interfere" with itself? These puzzles arise in the context of the classical physics view of reality, and the classical physics view grew out of the every-day common-sense view of physical reality, in which we are accustomed to regarding parts of reality such as rocks and trees, people and automobiles, baseballs and bullets, etc., as separate, discrete physical objects. We know now that, strictly speaking, this is not true, but in the world of superficial appearances, on the scale of the size of the objects that we deal with every day, we can operate reasonably well as if it were.

As we pursue a deeper understanding of the nature of reality, the wholeness and interconnectedness of all things becomes more and more apparent. Because of the extreme range of scale of the structures of the universe, this interconnectedness is not very apparent on the level of human perceptions, but becomes quite apparent at both the cosmological and quantum ends of the scale. Both relativity and quantum mechanics have reflected the underlying unity of reality in different and somewhat limited ways. Relativity had to account for the interconnectedness of matter, energy, time, space, and, to a lesser degree, the observer, in order to deal with relative motion at near light speed and phenomena on the cosmological scale. Quantum mechanics, on the other hand, has <u>forced</u> physicists to recognize the fact that, on the micro-microscopic scale of the quantum, the elementary particles of the atomic and sub-atomic realm cannot be considered to be objects apart from the apparatus of observation, or, ultimately, from the consciousness of the observer. In spite of starting with the assumptions of separateness and locality, and pursuing a reductionist approach, physical science has produced findings that clearly establish the interconnectedness of all things.

Now let's return to the double-slit experiment and see how transcendental physics brings the whole picture together: Within the framework of three-dimensional time, all of the photon states, represented by the Schrödinger wave equation as "possible" states, actually exist between the source and the receptors. In this broader, 3-D-time perspective, wave interference and particle phenomena are complementary aspects of reality and both occur between the source and the receptors. The phenomenon that an observer registers in his or her four-dimensional time-space universe, depends entirely upon the choice that he or she makes and the action taken (such as opening or closing a slit or blind) as a result of that decision. The reason that Wheeler and Cramer could conclude that consciousness and the observer play no "special" role in this experiment is because they have isolated the mechanics of the experiment from all other events, in effect, taking it out of context.

If we look at the whole picture, even just within a single observer's universe, i.e., leaving parallel universes and transactional offer-confirmation waves in 3-D time aside for the moment, we can see the error of the separation of consciousness and the observer from the phenomena of the experiment: While it is clear that an existing physical structure can act as the receptor for a photon, the elementary particles of that structure had to have their receptors, and while the universe has a large number of existing structures, if relativity and quantum mechanics are even partially correct, each and every quantum of matter and energy in all physical structures have to have had prior existing receptors. Without non-quantum receptors, we have a situation that leads to catastrophic contradiction by infinite descent. And the evidence that non-quantum receptors exist in consciousness is overwhelming. The most immediate evidence is your own consciousness. Without non-quantum receptors, communication would not be possible. The information bits conveyed by photons from the printed page would either be absorbed somewhere in inert, uncomprehending matter, or would have to find additional quantum receptors in ever-increasing numbers, causing the brain of the receiving sentient being to grow like a mushroom.

Within the total context of reality, both of the outcomes discussed in the double-slit experiment, *and many others*, actually exist. All of them originate as logical patterns in the substrate of reality and finally return to it, via strings of source/receptors, when the loop is completed by a non-quantum conscious receptor. The interference patterns build up on the photographic plate because the timeline that we create by separating the individual photon releases does not exist from the photon's perspective. The relativistic effect (expressed by the Lorentz contraction equation) requires that, from the photon's perspective, travelling at the speed of light, the space of our universe is collapsed to a point along the line from source to receptor. Therefore, as soon as the photon is released from the source, the wave of all possible photon states exists at the source, slits, and the receptors, simultaneously. When a conscious observer makes a choice and acts upon it, a receptor registers the effect of a photon (one of the photon's possible states), and the observer becomes aware of an addition to the wave interference pattern, or a click of one of the photon counters, depending on which choice he made.

And so we see that the puzzles associated with this experiment arise because of only dealing with a fragment or portion of reality, selected from the multi-dimensional continuum of total reality by the limited act of observation. We can decide whether the photon behaves as a particle or a wave in our universe after it passes the slits because it does not exist as a physical object, either particle or wave, until it registers as an impact on a receptor in a four-dimensional universe. And we find the wave interference patterns puzzling only if we believe that our experience of time is both complete and absolute.

The disagreement between Einstein and Bohr began over the interpretation of the two-slit experiment and reached a stalemate with the EPR paradox. Recall that Einstein and associates devised the EPR paradox as a means to disprove Heisenberg's uncertainty principle. In their conception of local reality, all of the information about a particle could be gained by measuring its twin. In fact, the EPR proposition was a very clever set up. Einstein knew what Bohr and Heisenberg's answer would be: Elementary particles don't exist until they register by impacting a receptor. Or, as Wheeler puts it, "No elementary phenomenon is a phenomenon until it is a registered phenomenon". But this means that when one twin is intercepted and forced to reveal its spin (in the case of an electron) or its polarization (in the case of a photon) the other twin must instantaneously exhibit the corresponding physical characteristic (spin or polarization). And thus, the twin particles are somehow connected, no matter how far apart they may be, even on opposite sides of the universe. This, Einstein thought, would be patently and obviously ridiculous to everyone. And from the viewpoint of a four-dimensional space-time universe, he was right. But the EPR paradox is a paradox only in a local, 3-D space, 1-D time, universe. The Aspect experiment proved that reality is not that simple.

Elementary particles, such as electrons and photons, appearing to move at light speed in the normal space time of human perception, are simply part of the fabric of the multi-dimensional space and 3-dimensional time of total reality. An elementary particle casts its shadow in every possible parallel universe. When caused to manifest in any one of them, it traces a timeline in that universe. All of the timelines from all possible universes, combined with all the energy and matter from all possible manifestations of elementary particles, make up the fabric of reality, the extent of which encompasses an infinity of possible universes in three-dimensional time. From the perspective of this total reality, the EPR paradox is not a paradox, and the puzzles of quantum mechanics are not puzzles. And the perspectives of relativity and quantum mechanics are not mutually exclusive, they are complementary.

Of course, transcendental physics is not the first theory to claim to resolve the EPR paradox and explain quantum observations. Indeed, Einstein did not consider the EPR thought experiment a paradox at all. He considered quantum theory to be an incomplete and inadequate description of reality. And it seems there are as many explanations and interpretations of quantum mechanics these days as there are physicists. But transcendental physics has a lot more to recommend it. It uses the language of distinctions, a mathematical system first derived by G. Spencer Brown, which is more basic than the mathematical systems being used and logically superior to them, especially for incorporating the existence and functioning of consciousness into the description of reality. This mathematical system goes beyond the normal binary logic of boolean algebra, which includes only true, false, and meaningless statements. The calculus of distinctions allows the inclusion of complementary aspects of reality, which when taken out of context may appear contradictory.

The explanations of transcendental physics do not depend upon reductionism or the isolation of a problem from the rest of the universe, as so many physical explanations do. On the contrary, the concepts of transcendental physics integrate the complementary aspects of reality that have been revealed by relativity, quantum mechanics and other fields of endeavor. Transcendental physics provides a framework that will allow us to integrate existing knowledge and the various legitimate efforts of individualized consciousness to understand the nature of reality and our relationship to it.

Returning to the reality of multiple-timeline universes, an important question arises: How does the conservation of matter and energy operate in a multi-universe reality? Are matter and energy conserved in each individual universe, but multiplied without limit as additional universes are brought out of the universal wave equation by conscious observations? If a universe, once it has been drawn out of the universal continuum by observation, exists totally independent of all other existing or potential universes, then the question of over-all conservation of matter and energy has no relevance within individual universes. But there is strong evidence from both relativity and quantum mechanics that there are numerous existing timeline universes and that they are not totally independent. Not only are they connected through primary consciousness, individual consciousness may, under some circumstances, have access to more than one of them, and there may be physical connections, as well.

In a reality where all physical manifestations are derived from the substance of infinitely continuous primary consciousness, the ultimate source of matter and energy is infinite. It follows that the conservation of matter and energy holds only within a closed physical system. By moving from one timeline universe to another, it may be possible to add energy and/or matter to a given universe. In this way, it may be possible to tap the infinite energy shown to exist in "empty" space by the Casimir experiment mentioned in Chapter 5.

How does the empty space of one timeline universe relate to the space of another? What are the possible physical connections? The equations of the general theory of relativity predict that the curvature of space- time can become so severe around any mass whose density exceeds about three hundred earth masses in a space the size of a golf ball, that no matter or energy can escape from its gravitational well. Such an object has been called a 'black hole' since not even light can escape. As long ago as the early 1920s it was shown that some stars are massive enough to eventually collapse under their own weight and form black holes. And now, within the last year or so, astronomers using the Hubble space telescope have announced the discovery of objects that have the characteristics of black holes. Mathematical analysis of black holes by Einstein and Rosen around 1935 indicated that black holes might actually be tubes connecting parallel universes. Whether or not such gateways between universes can ever be used by conscious, living beings remains to be seen. If access to other timeline universes through physical structures is problematic, what about gateways in consciousness? The Aspect experiment has shown reality to be nonlocal, and we have discussed evidence strongly suggesting that reality and what we've been calling primary consciousness are one and the same. It is not an unreasonable hypothesis that, through the nonlocality of consciousness, individual consciousness might be able to experience more than one universe. To test such an hypothesis, we have to accomplish two difficult, but not impossible things: First, we have to rid ourselves of the limiting belief that the individual consciousness which we experience is nothing more than the product of a physical body. Second, and just as important, we have to determine exactly how to achieve real inner objectivity that will lead to repeatable, verifiable, conscious experience. These necessary tasks define a very promising approach to an important area of research for transcendental physics in the near future: the objective determination of the mechanics of the interaction of individual consciousness with physical reality.

The recognition of the existence of non-quantum receptors associated with the bodies of sentient beings is the first step in accomplishing the first task. The next step is to investigate the non-quantum aspect of individual consciousness to determine whether, and to what extent it may function independent of the body. There is a considerable body of literature discussing claims of out-of-body episodes related to near -death experiences, drug-induced trips, and religious experiences. Transcendental physics provides a theoretical framework and a mathematical tool, the Calculus of Distinctions, that may be used to investigate such claims.

Approaches to the second task, achieving inner objectivity, might be developed by studying the techniques that have been used for centuries by Christian, yoga, zen, and Sufi meditation practitioners, and relating them to what we learn from our investigations in the first task. As suggested in Chapter 7, acquiring the ability of seeing the inner light of consciousness in the nexus between individual and primary consciousness should enhance the ability of the investigator to objectively discern the archetypal patterns of primary consciousness.

The success of transcendental physics depends on the establishment of investigation procedures for a new paradigm. This new paradigm is radically different than the current theories of scientific materialism from which science and technology have operated for so long. Resolution of the Einstein-Bohr debate over the ultimate nature of reality has proved them both wrong. In the next chapter we will summarize the new paradigm and have a look at what it can do for us and where it will take us.

CHAPTER 10.
TRANSCENDENTAL PHYSICS
THE NEW PARADIGM

The Cosmic Whole

Bathed in diurnal rays both day and night,
We sleep --always blinded by the light.

What's new is old, what's old is new.
Once discerned, what's one is two.

The simple seed becomes a tree.
The atom's speed, --Eternity.

Both time and space
Enshroud the soul
'Til we embrace

The Cosmic
Whole.

And now, though many pages have been written concerning the physics that transcends physics, what has been communicated remains to be seen. Ultimately, there is only one source of light, but in this hall of mirrors we call the real world, there are many reflections and certainly some illusion. From one point of view, all that one sees and experiences is nothing more --or less-- than reality communicating itself to the observer. But the message received depends largely upon how much we are willing and able to accept. It seems that many scientists are content to use quantum mechanics and relativity to solve problems and improve material technology without being concerned about what the new findings mean or what they imply about the nature of reality. They prefer to leave questions concerning truth and the ultimate nature of reality to the philosophers. In their minds, physics and metaphysics do not mix, or if they do, someone else will have to deal with it. They are too busy doing physics, or biology, or chemistry. Their position is understandable. They have every right to limit the scope of their efforts to the areas for which they have been trained, and to those things which interest them. But for this author, physics without metaphysics is like a pie without filling. Physics without metaphysics is not really science, it is engineering under the guise of physics.

Don't misunderstand me, there is nothing wrong with engineering. Engineering is the necessary and practical side of science. Without engineering, the fruits of science would do no one any good. But those involved in the pragmatic application of known scientific principles to real life problems rarely have time to concern themselves with metaphysical questions. To those who build bridges or particle accelerators, metaphysical questions are good for discussions over a beer, or on other occasions for intellectual socializing, but stand like thorny thickets or quagmires along the road to getting their job done. Too much time in metaphysical sidetracks can lead to confusion -- or even insanity. It seems that dealing with metaphysics is a nasty job, but someone has to buckle down and do it.

When I was midway through high school, I was very interested in science and mathematics, but I just barely tolerated the other courses in the curriculum, including the language arts. As it turned out, I was fortunate to have at least one very wise teacher. Her name was Mrs. Roberts and she taught Junior and Senior English. She was also the Student Counsellor and Advisor.

"You may become the greatest scientist who ever lived," she said, "but if you cannot communicate what you've learned, it will do no one any good."

Mrs. Roberts fanned the fire of my desire to learn. She introduced me to the works of many great communicators, including Ralph Waldo Emerson and Henry David Thereau. I have recently learned that others among her former students also recognize the importance and magnitude of the positive impact Mrs. Roberts had upon the outcome of their lives. I only hope that some of them were able to thank her and let her know how much she mattered.

Why study science --or anything else? It is only human to desire to know, to delve deeper, climb higher, to go farther, than anyone has gone before. Anyone who truly applies him or herself wholeheartedly to the pursuit of knowledge and understanding is an explorer, no different than one who explores mountains, or jungles, or an unknown terrain of any kind. What motivates an explorer? Why put one's life and limb at risk to reach the top of

Everest, the bottom of the Marianas Trench, or the vacuum of outer space? We've all heard the answer "Because it's there". Yes, it's there. But there's much more to it than that. By going there, and seeing, experiencing, feeling what it is to be there, we add something immeasurable to ourselves; we grow in knowledge and understanding.

There is the anticipation of the unknown, the thrill of discovery, and the feeling of accomplishment that comes from knowing that you have gone where on one has gone before. But above and beyond this, an important process is occurring, a process of integration: the experience, knowledge, and understanding is becoming a part of one's being.

Take, for example, the exploration of a cave. When you slip down a rope or climb over a jumble of rocks into the mouth of a cave, you feel the cool breath of another world, an unknown world, a world of blackness, presenting a stark contrast with the usually green and sunlit world that exists outside the cave's entrance. The cave is a world of radically different sights and sounds, different forms of life, a world of endless night. The explorer is aware of the unique beauty of each new vista, revealed perhaps for the first time by his flashlight. He is aware of the dangers of the slippery rock and seemingly bottomless pits that may swallow him up, or even snuff out his very life. He discovers delicate rock formations like crystalline flowers and billowing stone curtains, white crystal rimstone pools containing water so clear that it only becomes visible when its surface is disturbed.

The experience of each moment and each detail is exhilarating. And the details are additive so that at the end, when the cave has been explored, the knowledge of the cave's twists and turns, its formations and its life forms, have been integrated into an over-all picture: a map. The cave, its extent and features can then be put into context. They can be understood in relationship to the total landscape, the geology, surface topology, hydrology, and biology of the region.

In this book we have attempted to explore the mysterious territory beyond the quantum, where mind and matter meet. Our pursuit of understanding through transcendental physics may be valuable in terms of exploration for exploration's sake, but where does it take us? What do we gain by looking at reality in this way? Are the parallel universes and backward-moving energy waves of three-dimensional time that explain quantum and relativistic observations real? If so, how can we benefit from knowledge of them?

We know that reality is more than it appears to be. As Woody Allen is reported to have said:

"There is no question that there is an unseen world. The problem is how far is it from midtown and how late is it open?"

Our trip down the River of the Search for Reality started with a look back at the mighty tributaries of relativity and quantum mechanics and the giants who rode their daunting currents in crafts of their own design. At their confluence, waves clashed and the relatively clear waters of one seemed loathe to mix with the murky probabilistic waters of the other. But after a dizzying ride on the whirlpools of the Maxwell - Schrödinger wave equations, we were swept through the dragon's teeth of the double slit and over the Participatory Falls in the objective barrel of individual consciousness, in mortal fear of losing both our past and future to the smokey dragon of Copenhagen and the brutal rocks of Aspect. But our fears were foolish and unfounded, for now we find ourselves swimming gracefully in the calm waters of a new paradigm, free of the confining constraints of the barrel, enthralled to see the strange beauty of the exotic flora and fauna lining the banks of the river as it flows through the exciting new territory of this re-integrated universe of matter and Mind.

But this has been little more than a travelogue. The real purpose of transcendental physics is to prompt you to make the real journey, in a craft of your own design. Whether clinging to a crude raft of logs bound together with the strands of individual hopes and fears, or sitting on the proud deck of the yacht of scientific materialism, we are all being swept toward the falls. Whatever the cut of the jib or slant of the bow, the planks of belief in our decks and hulls will be split asunder by the truth of the falls. We may be loath to leave the warmth of our cabin or barrel, like the bird accustomed to the cage, reluctant to fly, even when the door is wide open. But the freedom of an infinite reality beckons.

Like the proponents of determinism and probabilism, some of us may be hesitant to mix the rich, warm flow of personal spiritual discipline with the crystal waters of physical science. But we must remind ourselves of the goal that we set when we embarked upon the journey: It was to know and understand the nature of reality in its totality, not just a limited aspect of it. Our goal was to know the ultimate nature of reality, the physics, chemistry, biology, and spirit of it, including even the meaning of life and existence. All the better, if we manage to enjoy the many interesting sights and experiences along the way, but surely no one wants to be stuck forever in a stagnant backwater of thought. At some point, we long to break free of the moorings, even if they are of our own design, and push out into the current that will carry us at last to the Sea of Enlightenment.

The way has been mapped by those who have gone before, both materialists and mystics. But those who would have you linger with them in the backwaters see no compatibility in the maps. They believe that the inner and outer worlds are light years apart. For them, Captain Jean Luc Piccard will never meet Jesus of Nazareth; the Zen contemplative never become a physicist; the cosmos never fit within the mind of man. But now and then, a bright light flashes and we see the maps merge, the paths converge, and the goal comes into view. Yet, in our ordinary day-to-day lives, we still seem to be either in the cabin or in the barrel.

We should not be content to remain ignorant of the vast spectrum of reality, blind to everything except a narrow band of perception. The human soul wants to know the meaning and purpose of all that we experience. Why do we experience pleasure and pain, ecstasy and suffering? We want to be aware, but even more than that, we want to understand the miracle of awareness itself.

Remember the story about the group of blind men who had no knowledge of elephants? After they were each allowed to feel one part of a live elephant, (one felt the trunk, one an ear, one a leg, and so on) they gave hilariously disparate descriptions of the animal. Well, with respect to trying to determine the nature of reality, we human beings are in the position of one blind man. We get to feel only the tail,--or maybe only a hair on the tail. Not only that, now that we know that there is more to the beast, we find that the elephant is a great smokey dragon and our part of it doesn't even exist until we feel of it!

We are in the position of one blind man because we all have roughly the same limitations placed on our observations. There are no other "blind" men, at least not that we know of, with whom to compare notes to form a composite picture of reality. But the remarkable chain of events from the EPR paradox to the Aspect experiment has opened the door a crack to give us a glimpse of the whole picture. This glimpse tells us that we have only seen the tip of the iceberg. Even the probings of physical science into the atom and the elementary particles may well have been trivial and peripheral, related to what waits for us in the complex whole of reality.

In "A Note on the Mathematical Approach" at the beginning of Laws of Form, George Spencer Brown remarks that "...mathematical texts generally begin the story somewhere in the middle, leaving the reader to pick up the thread as best he can. Here the story is traced from the beginning." Similarly, physics and all of the existing disciplines of physical science, start somewhere in the middle, in their efforts to understand the nature of reality. As with Laws of Form, the approach of transcendental physics is to trace the story from the very beginning, i.e., with the drawing of the first distinction. What is the nature of reality? With what we have learned in our pursuit of transcendental physics, we are now ready to answer that question. Let's summarize what we've found:

In the first chapter, we compared the course of scientific thought to the flow of a river, with quantum mechanics and relativity as major tributaries. We discussed the logical conflict between relativity and quantum theory, and examined the Einstein-Bohr debate, the EPR paradox, Bell's theorem, the Copenhagen interpretation, and the Aspect experiment. We saw that the results of the Aspect experiment validated the Copenhagen interpretation and revealed nonlocality as an underlying feature of reality. We observed the fact that while most scientists have accepted the results of the Aspect experiment, few understand what it means. Finally, we discussed how these results cry out for a new paradigm that will include consciousness as an integral and interacting part of reality.

In Chapter 2 we noted the resistance of the scientific establishment to new paradigms, the reasons for the resistance, and the prevalence of the emergence theory of consciousness in the current paradigm. We discussed the two-slit and delayed-choice experiments, we proved that consciousness has to function as the final receptor of free quanta of matter and energy, and that some form of consciousness had to precede the first quantum in the creation of the universe. We called that form of consciousness "primary". We found that the substance of consciousness could not be composed of quanta of matter and energy, and that it was nonlocal and capable of containing complex structure. We pointed out that the proof of the necessity of non-quantum receptors generally reverses the assumptions of scientific materialism, especially the emergence of consciousness theory.

In Chapter 3 we saw that consciousness cannot be explained in terms of matter, while matter, on the other hand, can be adequately explained and understood in terms of consciousness.

In Chapter 4 we examined the metaphysical depth of the thinking of some of the outstanding contributors to the advancement of science and contrasted their spiritual grounding with the shallow materialism so prevalent today. We discussed the need for meaning and the possibility of the evolution or expansion of individual consciousness to higher states, perhaps even toward the all-encompassing knowledge of primary consciousness.

In Chapter 5, we found that current physical theories, including relativity and quantum mechanics, are inadequate to explain the complex phenomenon of light, and that observation as we know it precipitates only one aspect of the "smokey dragon" of Schrödinger's wave equation. Reality is whole, not fragmented, and the final step in the chain of events in any observation, the *receptorium* and *gestaltenraum*, or *nexus* of consciousness and reality is, at least, the seventh in the sets of quanta and receptors.

Then, in Chapter 6, we identified the drawing of distinctions and the organization of distinctions into structure as the primary and secondary functions of consciousness, and we discussed the role of consciousness in the creation of the structure of physical reality. We saw that time, space, matter and energy as we experience them, can only exist and be described from the point of view of a conscious observer.

In Chapter 7, we discussed the possibility of and need for objectivity in the study of consciousness, we briefly reviewed G. Spencer Brown's Laws of Form and the need for, the derivation of, and the application of the Calculus of Distinctions. We demonstrated that the common structures and forms found in the physical universe pre-exist in consciousness and may be studied as structure independent of the physical world, using the Calculus of Distinctions. We found that the Calculus of Distinctions may be used to distinguish between perceptual, conceptual, and existential distinctions, and that Reality is self-referential.

In Chapter 8, we pursued the possibility of the objective investigation of consciousness by examining methods and techniques for refining individual consciousness for use as a tool of scientific inquiry. We focused on the *receptorium* as a doorway into primary consciousness, and speculated about what we might find there. In the end, we found that there is strong evidence that consciousness and energy, mind and matter, man and the universe, God and the cosmos, all are part and parcel of one whole thing, the warp and woof of the same cloth, ultimately one and the same.

In Chapter 9, we saw that the opposites and extremes that we perceive as individual sentient beings converge in primary consciousness. The reality we perceive contains the appearance of discreetness, opposites and extremes only because it is a partial view of reality selected for us by the limitations of our apparatus of perception. The whole of reality is nonlocal and infinitely continuous. The appearance of discreet particles is caused by individual acts of observation that precipitate them from the underlying continuity of the substance of primary consciousness. Using the Calculus of Distinctions to distinguish between perceptual, conceptual and actual or existential distinctions, we discovered that time must be three-dimensional. Three-dimensional time enables us to understand nonlocality, explains relativistic and quantum observations, and resolves their conflicts. Thus transcendental physics integrates relativity and quantum mechanics and emerges as a viable theoretical basis for a new scientific paradigm. Not only does it show us how to integrate the main branches of physical science, but it also provides a natural basis and starting point for the integration of all aspects of the search for truth.

So what does this all mean? Where do these discussions, proofs, and observations lead us? Exactly how does transcendental physics answer the eternal questions of the nature of reality? Let's list and clarify what we know and see what questions remain.

Profile of the New Paradigm. The new paradigm of Transcendental Physics is based on:

1. The fact that free quanta of matter and energy have no paths or separate existence until they register on a receptor. (The Copenhagen interpretation of quantum mechanics)

2. Bell's theorem and the results of the Aspect experiment prove that the Copenhagen interpretation is correct and that reality is nonlocal.

3. The logic of infinite descent proves that non-quantum receptors exist and that a primary form of consciousness had to exist before the appearance of the physical aspects of the universe.

The following concepts are derived from and consistent with these facts.

A. Separate, localized quanta of matter and energy, and the parameters of extension: space and time, are produced by the functioning of consciousness.

B. The primary and secondary functions of consciousness are: the drawing of distinctions and the organization of distinctions into structure and form.

C. Consciousness functions as a non-quantum receptor, completing the loop of consciousness observing itself.

D. Individualized consciousness operates over a limited range and participates in the formation and perpetuation of structure and order in the universe.

E. Conscious observation selects a limited aspect of reality from the continuum of possible states due to the limitations of the apparatus of observation.

F. Time, like space, is three-dimensional but the observations we make as individual conscious observers limit our experience to events that define a single timeline.

G. All order and structure manifested in the universe originate in primary consciousness.

H. Primary consciousness, and therefore, reality is nonlocal, participatory, and self-referential.

I. There are two types of consciousness: primary and individualized.

J. Consciousness and matter/energy are complementary aspects of an infinitely continuous reality.

Within this framework, it becomes clear that a complete and consistent science may be developed by beginning with the first distinction. In order for there to have been a first distinction, there had to have been a primary consciousness. But what is primary consciousness? How does it differ from the individual consciousness that you and I enjoy?

We note that consciousness is a subtle, interactive form of the universal substance from which all matter and energy are composed. The Aspect experiment proved that reality has nonlocal features. Since any break in continuity would constitute a boundary, and thus a distinction, primary consciousness, existing prior to any distinction, would have to be completely nonlocal and infinitely continuous. The first distinction, therefore, had to be drawn *within* and *by* primary consciousness itself. By tracing the process of the conscious forming of distinctions within our own individual consciousness back to the first or primary distinction, we find that prior to that distinction, there were no boundaries and thus no resistance to the formation of the first distinction. This lack of resistance continues to exist at the edges of the universe and explains why it is expanding. But in order to prevent the universe from expanding as an infinitely continuous wave form forever, selected parts of the innate structure of primary consciousness were projected to act as receptors that would bring the corresponding structures out of the infinitely continuous spectrum of primary consciousness. These localized distinctions became elementary particles in the time and space that their structure defined, and provided the building blocks for the externalization of the innate structure of primary consciousness as the physical universe. As sentient beings in this universe, we possess both corporeal form and individualized consciousness, and our function is to participate in the flow of the structure, form, meaning, and consciousness from primary consciousness into finite, physical manifestation.

As G. Spencer Brown observed, and we demonstrated in Chapter 6, once the first distinction is made, all of the forms of the universe follow. The forms and structures of the universe have therefore evolved as a result of the act of the formation of the first distinction. Furthermore, without the continued operation of consciousness, the second law of thermodynamics, known to operate in all closed mechanical systems, would bring about the rapid, if not instantaneous, dissolution of the physical universe. Therefore, the continuing functioning of primary consciousness is ocurring on the quantum and cosmological scales for the perpetuation of the physical manifestation of form and structure.

Why are we not normally aware of the functioning of primary consciousness? Because of the relative size of the atoms, cells, and molecules that make up our sense organs and apparatus of observation and perception. Since primary consciousness contains all possible form and structure, it is intelligent and purposeful. It is no accident that the size of the cells making up the bodies and sense organs of sentient organic entities such as human beings, is billions of times larger than the individual distinctions that create and sustain the physical universe. This size difference limits the range of wave lengths of energy that may be utilized to perceive the universe and conceals the rest of reality/primary consciousness from our direct awareness. The so-called universal physical constants $\mathbf{c}$ and $\mathbf{h}$ are functions of this size difference. The particular values of these constants in our universe are numbers determined precisely by this size ratio. They are universal from our point of view as human beings, but would have different values for sentient beings whose physical universe might be composed of larger or smaller elementary units.

We've discussed mind, structure, and matter, focusing on the interface of our individual consciousness with the physical world in an effort to further our understanding of the nature of reality. Bell's theorem, the Aspect experiment, and the proof of the involvement of consciousness by the infinite descent of receptors have led us to understand that the universe is comprised of logical structure flowing from primary consciousness. With conscious observation, concrete forms are selected from the infinite continuum of possible forms, and the self-referential loop of consciousness observing itself is complete.

As individual manifestations of consciousness, we are naturally concerned with knowing who and what we are, the meaning of life, and the significance and purpose of our existence. In General, human life and all other life forms, the evolution of the physical universe, and the relationship between living and inanimate matter can be brought into proper perspective by thinking in terms of energy. The energy utilized by individual observers to form and manipulate images made up of individual distinctions is quantified in light as photons, in electrical energy as electrons, and finally in mechanical energy as mass in motion, or momentum, meted out in multiples of $\hbar$, Planck's constant.

A complete view of reality cannot be gained from the viewpoint of a single observer, but can be conceived of as being composed of, or built up by, the integration of the views of all possible observers. In this integrated view, there is no time or space, since all possible observers fill all time and space. Thus we see that relativity and quantum mechanics, each successful in predicting certain specific aspects of reality, provide only a partial and incomplete description of reality. To complete the description, we must add the substance and functioning of primary consciousness which integrates individual observations into nonlocal, infinitely continuous reality.

We have demonstrated that primary consciousness had to exist prior to the manifestation of any physical form, and that it must continue to function at the quantum level in order to perpetuate the existence of the physical universe. Then, because of the limitations of finite human perception, the following question may arise: If primary consciousness is pervasive throughout the universe, why doesn't it cause the collapse of the quantum wave equation of possible forms, negating the need for individualized consciousness? The answer is that primary consciousness, being universally nonlocal, is aware of the spectrum of *all* possible forms imbedded within the universal wave function. However, the manifestation of all possible forms at once would result in an infinitely continuous form no different than primary consciousness itself, while the finite form selected by the observation of an individual sentient being is caused to appear distinct and separate from the matrix of primary consciousness by the act of finite observation. The observation of a finite number of distinctions reveals only a finite portion of reality.

With the description of the new transcendental paradigm of transcendental physics more or less complete, we may now ask: Of what practical value is such a paradigm? The most obvious response is that it is eminently practical in that it provides a theoretical basis for uniting all the different forms of our search for truth. The recognition of consciousness as an integral part of reality, and the proof of the necessity of the existence of consciousness prior to the formation of physical reality, restores meaning to existence, and purpose to life in the physical universe. We know now that we live in a universe where mind and physical form are linked through the mechanics of observation. This knowledge will allow us to develop a rational, scientific approach to the study of consciousness and to integrate the sciences of consciousness with the physical sciences.

For an example of the integrative power of transcendental physics, we may turn to the problem of integrating relativity and quantum mechanics. While it is beyond the scope of the present book to derive unified field equations from the Calculus of Distinctions, we can demonstrate that it is possible, at least in principle. The reason that neither Einstein, nor any physicist working within the current paradigm, has been able to complete the mathematical formulation of a unified field or grand unification theory is because their basic assumptions leave out an essential part of reality. They do not include the substance and functioning of consciousness. And the mathematical tools currently in use are not effective in describingg the interaction of consciousness with energy and matter.

The necessity of the action of consciousness in the formation and propagation of physical phenomena, demonstrated in the proof of the infinite descent of receptors (Appendix C) clearly points up these deficiencies in current theory. In order to include consciousness and its interaction with physical reality, a general expression may be written in the Calculus of Distinctions including consciousness, energy, and matter as three different parameters describing the basic substance of reality, along with the pair of three-dimensional variables of extent: time and space. Transformations from one parameter to another may then be derived using the principles of relativity, quantum mechanics, and the primary and secondary functions of consciousness, i.e., the drawing and organizing of distinctions. Equations describing relativistic, quantum, and logical structure transformations will be found to be imbedded in this general Calculus of Distinctions expression.

There is no question that we need a consistent paradigm, one that embraces all legitimate methods of searching for truth, and transcendental physics offers that possibility. But real science, however far-fetched it may have seemed in the beginning, has always led to real benefits. Relativity, thought at first to be not only beyond the comprehension of most people, but also irrelevant as far as most practical applications were concerned, has found numerous applications and has profoundly changed our lives. Quantum mechanics, with all its weirdness of discontinuous 'quantum jumps' and 'spooky' instantaneous, nonlocal action at a distance, has predicted phenomena that were never suspected, and which have led to many useful technological advances. Is there a similar potential for technological benefits from transcendental physics?

The proponents of the anthropic principle have long contended that consciousness plays an important role in the universe. But they had no scientific framework within which their contention might be tested. In the name of objectivity, science embraced the assumption that consciousness plays no role at all in the evolution of the physical universe. The proof that consciousness is a necessary ingredient in the formation and organization of the physical features of the universe is the foundation of transcendental physics.

The inclusion of consciousness in the mathematical description of phenomena is accomplished by means of the Calculus of Distinctions, which gives rise to the proof of the necessity of a non-quantum substance. The technology of the future may be able to take advantage of the knowledge that four-dimensional space-time is an artifact of individual consciousness to develop methods of distant viewing in space and time by means of the nonlocality of three-dimensional time.

Transcendental physics transforms the study of consciousness by`providing a solid scientific basis for explaining the interaction of mind and matter. The recognition of the existence of a conscious, non-physical *receptorium* within the physical vehicle of each living organism, functioning as a link between consciousness and matter, provides a new approach to understanding and explaining the nature of consciousness. Scientists of the future can be trained to develop inner objectivity, which will enhance their ability to make direct observations of the flow of form and structure from the primary source of order into the physical universe.

The knowledge that all of the forms that we perceive are manifestations of the innate structure of infinitely continuous primary consciousness, linked to our individualized consciousness by its universal nonlocality, requires that the basic materialism of modern science be discarded in favor of a new transcendental ontology. Once it is understood that reality is much more than matter and energy interacting in time and space, and that this greater reality can be investigated objectively, the doors will be thrown open for science to grow as never before. Science is poised on the brink of discovering a whole new universe that exists beyond time and space. That greater reality has, of course, always been there. Strangely, the barrier has not been one of insufficient capability or opportunity, but one of disbelief. But now, the discovery of empirical evidence and mathematical proof have dispelled disbelief.

Proof that the final receptor of all consciousness, either individualized or primary, must be nonphysical, implies that all physical forms, from galaxies to biological organisms, are vehicles utilized by consciousness to project form and structure into the universe and are evolved for that purpose. Thus all existence has meaning and purpose. The nonlocal nature of reality revealed by Bell's theorem and the Aspect experiment proves that all forms of structure, from the particle-wave structure of the hydrogen atom to the most complex organic structure in the universe, are like blossoms on a single tree, manifestations of the whole. When these truths become generally known and accepted, the world will change. There is no question this knowledge has the power to revitalize science and human society.

The new science of transcendental physics integrates the search for truth. It provides a natural framework for the investigation of matter, energy and consciousness. It holds forth the possibility of uniting the reality of consciousness with the reality of the physical universe. If successful, it will allow us to understand, at last, who we are and exactly what our place is in the universe.

CHAPTER 11. NON-QUANTUM REALITY

We've seen the rainbow beyond the falls. We know that interesting and strange new worlds await our exploration, but just where will our search take us from here? Transcendental physics is only the first step toward a comprehensive transcendental science. The most important result of the creation of a transcendental scientific paradigm will be its power to integrate all aspects of the scientific search for the nature of truth and reality. But exactly what is the nature of non-quantum reality? Is primary consciousness the God of Christianity and/or other organized religions of the world? Or is it something different? Will the exploration of the connection between mind and matter, between individualized consciousness and primary consciousness re-establish our belief in, and knowledge of God, or will it bring a revolutionary new understanding of the universe and our relationship to it? The scientific study of non-quantum reality will surely expand our knowledge and understanding, but what will this expanded knowledge and understanding bring?

Optimistic thinkers of all times have believed that ignorance is bondage and that true knowledge frees body, mind, and soul for higher purpose. How will transcendental science affect our lives? Transcendental science has the potential of shattering the illusion of a closed, dead materialistic reality that has been fostered by the current scientific paradigm, bringing new meaning and purpose to life, and opening the doors to the wonders of a transcendental reality that we've only dreamed of before.

How do we go about exploring non-quantum reality? The question of how to know the nature of reality has been, and will always be, a difficult one. Even in the beginning of the Age of Science, it was recognized that reality was a vast, complex and mysterious thing. Just how vast and complex it is was not known, but it seemed very unlikely that anyone could understand it all in one lifetime. Because of this daunting prospect, early scientists focused on those parts of reality that appeared to be the easiest to deal with: those things that could be separated from all the rest - things that could be weighed and measured - and physical science was born.

The science of physics was very successful. It seemed that physical objects behaved and interacted with each other in ways that were logical and predictable. Also, the elements of reality chosen for study constituted a realm that seemed to be pretty much self-contained. Standards and units of weights and measures could be defined in terms of these same elements of reality and the scientist could stand aside as an uninvolved observer - a very manageable arrangement. But this approach could only take us so far.

Physicists are, of course, human beings and, like most people, tend to think of everything in the terms they know best. Ask a shoemaker what the most important thing is for health, happiness, success, and satisfaction in life, and he might answer: "A good, strong, well-made pair of shoes." Ask a carpenter what's important, and you're likely to hear analogies concerning things like laying firm foundations, carefully cutting and fitting joists, studs, and rafters to properly support walls and roofs. It is only natural that physicists, successful in describing the basic mechanics of atoms, stars and stones, when thinking about the nature of reality, have forgotten that theirs is a very much over-simplified view, that many of the more subtle aspects of reality were excluded from their field of study in the very beginning. But the ultimate connectedness of all things has assured that we, physicists all at some level, have finally to come face to face with the existence of profoundly non-physical features in our search for reality.

Having established the necessity of non-quantum receptors, we must ask the following question: What is the nature of the non-quantum aspect of reality? How do we describe something that is neither matter nor energy, i.e., not composed of quanta? It has to be substantial, since it acts as the final receptor for the information carried by physical quanta, but it cannot itself be composed of quanta. From the standpoint of physics, we are introducing a new form of the substance of reality.

It may be helpful to think of the substance of reality as existing in four different forms defined in terms of varying levels of locality and density: 1.) Primary consciousness - universally nonlocal, continuous throughout all space and time. The source of the original non-quantum receptor that brought physical reality into being. 2.) Individualized consciousness - the substance of non-quantum receptors operating in a finite, bounded regions through all sentient beings. 3.) Energy - substance localized by the drawing of distinctions, causing vibration and wave forms. 4.) Matter - highly localized dense, structured forms reflecting the non-quantum structures existing in primary consciousness.

Using the various methods of classical physics, relativity, and quantum physics, over the past 150 years, scientists have probed the structure of reality, discovered and described matter and energy as quanta. We have now demonstrated the necessity of non-quantum receptors. Where is non-quantum reality and how does it relate to the phenomena of matter and energy? If consciousness is generalized as *mind*, in the sense usually attributed to Descartes and his followers, the question 'where is non-quantum reality?' seems to be similar to the question of the 'Cartesian cut': Where is the boundary between matter and mind? But we soon see that this is the <u>wrong</u> question. It is not a matter of size or location. Size and location are concepts of locality. The interface of consciousness with matter and energy is global because consciousness is nonlocal in nature.

This explains how consciousness can gain so much information from a quantum of matter or energy. First, the image or pattern constructed from the information content of quanta already exists in primary consciousness, which pervades all time and space. Second, individual consciousness, because of its nonlocal nature, is in contact with every detail of the quantum within the region of its receptorium.

Why do our observations reveal a world built up of quanta? Where is the evidence of non-quantum reality in our observations? In fact, our perceptions <u>are</u> non-quantum. The Copenhagen interpretation, verified by Bell and Aspect, asserts that we do not see quanta, only their effects. The images constructed in the gestaltenraum are not composed of quanta. They are not perfect parallel constructions with respect to the outer world. For perfect parallelism to exist, the image would have to be composed of quanta. For every quantum in the outer world, there would have to be a quantum in the image. This would require the brain of every observer to contain as many quanta as the outer world being observed. Since physical quanta are finite in size, the inside of our heads would have to be literally as big as all outdoors!

Consider a photograph. The image of a panoramic scene is captured on a few square inches of photographic paper. It is composed of quanta, but it is <u>not</u> a one-to-one copy of the panorama. There are many fewer quanta in the image than in the scene, and because of this, some detail is lost. In fact, it is this loss of detail that creates the illusion of distance and perspective in the photograph.

Approaching a tree in the actual scene, we find that every leaf exists in great detail. There may be birds, insects, spider webs, etc. that become visible as we look at the scene more closely. By closer and closer observation, we find that these details may be inspected right down to the cells, atoms and quanta that support their forms. But in the photograph, no matter how closely we look, there is only so much detail. Magnifying the photograph indefinitely, we eventually find the molecules, atoms, and quanta of the photographic paper, not the quanta of the trees, birds, etc. of the scene. The amount of detail in the photograph is inversely proportional to the ratio of the total area of the scene to the size of the paper.

But how far can we carry this analogy? The eye is, to some considerable extent, analogous to the camera; it has a lens, a focal point, and a photosensitive area, the retina, which is quite small in relation to the panoramic scenes it seems to capture. And what about the image perceived by the conscious observer? It differs from the photograph in a number of significant ways: It is not static, but is continually being updated by new bits of information and by internal organizing and processing. It appears not only to have perspective, but to be three-dimensional like the scene it portrays. As we approach the tree in the outer scene, new bits of information are continually added and incorporated in the inner image, revealing birds, spiders, etc., giving the illusion of a one-to-one correspondence with the scene. The internal processing will even supply details to fill in missing data, in order to present a complete image to the perceiving consciousness even when the data from the outer scene are not complete. The 'paving over' of the blind spot where the optic nerve is connected to the retina is a ready example. The consciousness of the observer selects and processes bits of information from a virtually infinite sea of data to construct its image of reality.

Where does the additional information used to complete the image come from? And how far will the internal processor go to complete an image as a logical pattern, acceptable to the perceiving consciousness? Could this explain why different people observing the same scene often report seeing different things? The additional information has to come from data already existing as logical patterns available to the consciousness of the observer. These patterns come from memory and/or projections of logical patterns from primary consciousness. This line of reasoning suggests picking up where G. Spencer Brown left off in the development of the logic of memory in Laws of Form (pages 61, 62, and 100). Applying the self-referential approach of the Calculus of Distinctions to the description of mental processes may prove to be a very fruitful area of research.

Attempts to explain conscious observations in terms of the mechanics of the physical and electro-chemical processes of the brain are bound to fail because they focus entirely on brain çells made up of quanta of matter and energy and overlook the necessary existence of the non-quantum receptor and the nonlocal functioning of consciousness. If all the various aspects of incoming data, such as the contrast of edges, depth perception, color, and intensity, are carried by quanta of matter and energy and received by the quanta of material brain cells, some mechanism must be found to explain how the organization and merging that must occur to form coherent images is managed by separate physical elements without requiring an almost infinite number of neural networks.

At the end of Chapter 3, we mentioned that proponents of mind-matter dualism, have proposed that the interaction between consciousness and matter takes place in a group of cells at the very top of the brain, a region known as the *supplementary motor area*, or SMA. This may well be the region containing most of the specialized cells that function to relay information to the non-quantum receptor. But because of the nonlocal nature of consciousness, the connection is global not local.

The awareness of a conscious entity can be directed to one or more specific regions of its body at will. If this were not so, we could not deliberately move our hands or walk across the room. People with specialized training, such as certain athletes and yogis, are able to become aware of subtle functions of the body such as breath, heartbeat, and the flow of blood, functions of which most of us are not normally consciously aware. It seems likely that a conscious entity is aware of and interacts with, on some level, every cell of his or her body. From the instant an information-bearing quantum impacts a sense organ, such as the eye of an observer, that bit of information is being relayed, processed and organized by the physical structures of the organs, nerves, and brain, and by the instinctive nonlocal action of consciousness. By the time the image is formed in the *gestaltenraum* and perceived by the consciousness of the observer, it is no longer composed of quanta of matter and energy, it is, like the final receptor, non-quantum in nature.

The recognition of the existence of non-quantum receptors as part of consciousness and the nonlocality of consciousness enables us to look at the questions about the nature of consciousness and the mind-body interaction problem from a new perspective. E.J. Squires, in '*Quantum Theory and the Need for Consciousness*', Journal of Consciousness Studies, (Vol No. 2, p. 202) says that we need "a calculation that will tell us whether a given physical system is conscious". This seems like a reasonable query, but one of the things that makes the derivation of such a calculation difficult, is that no physical system is totally separate from consciousness.

Order and structure originate in consciousness and flow from it into the physical universe, and all physical structures are encompassed and pervaded by primary consciousness. The very fact that a structure exists is evidence that consciousness has formed it and is sustaining it, since all form and structure is created, sustained, and destroyed by consciousness. Without the action of consciousness, the second law of thermodynamics would cause the negative entropy (form and structure) of physical systems to dissolve back into unmanifest primary consciousness. So, strictly speaking, all physical systems are manifestations of consciousness, and it might be more appropriate to seek a calculation that can tell us *to what degree* a physical system is conscious. Or, perhaps we need a calculation to tell us whether or not a system is *self-conscious*.

It may be possible to pursue this goal using the Calculus of Distinctions. At least in theory, any physical system, being finite, can be described using Calculus of Distinction expressions. Since memory can be discerned in such expressions (see Laws of Form, page 61 - 68), it is very likely that the level of physical complexity necessary for a system to be self conscious will be reflected in the level of self-reference or *subversion* appearing in the form of the expression describing it. All of the various functions of consciousness that manifest through the distinctions of quanta may be discernable in the form of the expression. Again, this appears to be a promising area for further research.

It is well enough for science to advance into a new

transcendental stage, but how will this advancement affect the average individual? Most of the people now living have been under the strong influence of an increasingly materialistic world view for most of their lives. This is reflected in the hopelessness, mental depression, and crime that is increasing in nearly every society on the planet today. The discoveries of science filter down slowly and the callous minds of those raised in an atmosphere of material greed and violence may be hard to reach. If we are rational beings, we must live our lives according to what we believe to be true and real. For the purpose of argument, let's consider two opposing views: materialism and transcendentalism.

If only material forms are real, we are nothing more than convoluted tubes of protoplasm, supported by a bony structure, capable of processing certain types of material for the purpose of extracting energy to maintain our bodies and replicate our species. A soft-tissue computing system directs our activities. When this protoplasm food processor finishes its natural cycle, or is destroyed by some catastrophic event, the whole system dies, degrades and ceases to exist.

If material forms arise from and because of consciousness, the body and all its functions are manifestations of form and structure originating in consciousness. In this case, physical bodies are vehicles evolved for the purpose of expressing and experiencing the structure and order of consciousness in physical form. When the body dies or is destroyed, consciousness is not affected or lessened. It continues to exist in its natural nonlocal form until the physical circumstances can be evolved once more to the point that it may function and express itself again in physical form.

Which view is correct? As we've seen, even physics

now favors the later. But let's look at it for a moment as if it were purely a matter of choice. If we choose to believe in materialism and conduct our lives as if only matter and energy exist, there are two possibilities: your belief will eventually prove to be either right or wrong. If you are right, when you are gone, it will not matter to you how you behaved while you were alive, since you will simply no longer exist. And, therefore, there is no reason not to be just as self-centered, cruel and violent as you wish. If, on the other hand, you are wrong, and your consciousness survives physical death, you will find that regret, guilt, and, in general, the fruits of your actions will still be with you in non-quantum reality.

If we choose to believe in the continuation of individual consciousness in non-quantum reality and conduct ourselves as if we are consciousness inhabiting a physical body temporarily, then there are also two possibilities. If you are right, continuity and cause and effect are operational and if your actions are beneficial to consciousness in general, i.e., to your larger self, you will grow and progress, to enhance your joy and bliss and to decrease pain and suffering.. If you are wrong, then, like in the case of the materialist, it won't matter to you once you're gone, but at least, some of those who remain may have reason to think kindly of you.

We have to conclude that, even if we could not prove the existence of non-quantum reality, the only logical and responsible choice is to behave as if there is a primary consciousness which is the source and substance of all reality. However, it is a point of fact that there is now no doubt that non-quantum reality does exist. To paraphrase Woody Allen, the really appropriate questions now are: Is it open to me, and how do I get there?

It is apparent that all physical systems reflect the basic

structures of logic, such as symmetry, reflection, progression, etc., originating in primary consciousness, and therefore, even the bodies and actions of sentient beings are governed by the over-riding principles of primary consciousness. However, we do have some degree of freedom to make conscious decisions that affect the course of our lives. The decisions we make may have considerable effect on other conscious entities and on the way our own consciousness grows, regresses, or stagnates over the course of a lifetime.

Clearly, we want the state of our consciousness to evolve toward increased knowledge and awareness of the infinitely continuous nonlocality of primary consciousness, not away from it. The more global or universal our nonlocal awareness, the more we realize the essential oneness of all things. With the complete realization of oneness, there could be no reason, no motivation to act in any way counter to the flow of the structure of primary consciousness into the universe. With the knowledge that what we feel at the center of our awareness is connected with, and even the same as, the center of being in every other individualized conscious being, all desire to act in a way that might harm or deter the greater awareness of any form of consciousness disappears.

Can we consciously evolve toward cosmic consciousness or oneness with primary consciousness, or receive it directly from the source, as our religious leaders have claimed? How do we get from the state of awareness in which we find ourselves, whatever that state may be, to such an elevated state of consciousness? The study of non-quantum reality should give us some clues; however, one thing is clearly necessary: we must be absolutely honest with ourselves. We must develop the absolute clarity of inner objectivity to the best of our ability. And care must be taken not to convince oneself that one has reached this state of consciousness before one actually has. Individual consciousness may become so convoluted that it can delude itself into believing that any self-centered action is for the greater good when, in fact, it may be just the opposite. The fact that this can and does happen is a result of the disparity between the freedom felt when individual consciousness comes in contact with non-quantum reality and the limitations and confinement experienced in the structure of physical reality.

How do we know when the delusions of individual consciousness have been dropped and oneness with primary consciousness has been realized? How do we recognize this state in others? Contact with such a person is contact with primary consciousness, and even a brief experience of primary consciousness will change your life. An individual immersed in primary consciousness *sees* the flow of conscious energy from the form and structure of primary consciousness into the physical universe, creating and sustaining it. Such a person is, as Bucke said: "...as far above self consciousness as is that above simple consciousness."

Reality is really very simple in both its inner and outer manifestations. It is only the endless machinations of our egos that obscure and complicate it, and thus, the importance of inner objectivity cannot be over emphasized. The mental confusion developed through years of ego-centered thinking dissolves when we are able to stop and watch it with total objectivity. When the inner dialogue subsides, and mental clarity arises, the ever-present background of primary consciousness comes to the fore. Once the light of reality shines into the mental mazes you have created to protect your imaginary ego, the clouds of erroneous thought disperse like morning fog on a sunny day.

The course of "natural science" has come full cycle. The pursuit of physics has yielded transcendental physics. How could this happen? It was bound to happen. Individual consciousness, applying itself with honest diligence to the study of any part of reality, even the lowliest clod of earth, not allowing itself to become entrapped by the logical imperatives of its own assumptions, must eventually come face to face with the Primary Conscious One, the essence of its own being, pervading all matter, energy, space, time, and consciousness. It is for this purpose that we were created.

DISTINCTIONS

Air mingles with new life's blood.
The first breath fuels a painful cry.
The last brings a memory flood,
and pale lips release a lifeless sigh.

We strain against the rigid frame
of time; we know we have to try
to learn to play life's daunting game,
to plumb the depths, to reach the sky.

But it's a lie that's told a million times.
These doors of pleasure and of pain
are only one. The door revolves,
as we pass through again, again ...

We can recall the endless throes
of me and you, and you and me:
Dewdrops sparkling on the Rose ...
Raindrops falling to the Sea.

APPENDICES

APPENDIX A

BELL'S THEOREM

Bell's theorem grew out of John Bell's study of the Einstein-Podolsky-Rosen (EPR) paradox, the pivotal argument of the Einstein-Bohr debate. The theorem points the way to the resolution of the dispute over the ultimate nature of reality. If Einstein's view is the correct one, elementary particles exist in the same way that baseballs exist between the pitcher's mound and home plate. They are physical objects existing locally, at every space-time point between source and receptor. If this is true, the EPR paradox (see Chapter 1) disproves Heisenberg's uncertainty principle, a basic underlying tenet of quantum mechanics. If, on the other hand, Bohr's view is correct, then elementary phenomena exist only as a range of possibilities, described by Schroedinger's probability wave equation, until they make impacts on receptors registering in some manner that can be recorded and/or observed by a conscious observer.

The logic and reasoning leading to Bell's inequality, or Bell's theorem, as it has come to be known, may be outlined as follows: Bell envisioned an experiment dealing with correlated particle pairs, similar to the EPR thought experiment, as described in Chapter 1. The particles comprising the pair are known to be 100% correlated at their source. Let's suppose that the particles are electrons. Electrons possess an intrinsic angular momentum, commonly called spin. This spin has been demonstrated (by the Stern-Gerlach experiment) to be quantized in both magnitude and direction, with only two orientations, that are equal and opposite. One is called "spin up", the other "spin down". Electrons are generated by certain types of sources in pairs in such a way that if one particle has "up" spin, the other always has "down" spin. Thus their correlation has a probability of 1. Since their spins are always opposite, the correlation is equal to -1. These electrons, however, are generated in a random, unpredictable manner, so that there is no way to know which electron will have "up" spin and which will have "down" spin.

The particles are projected in opposite directions, and are collected in spin detectors at some distance away from the source. See Figure 1, Chapter 1. If the spin detectors are rotated out of alignment, the probability of perfect correlation will change. If the particles exist in a local reality, as Einstein believed, we can't know exactly what will happen between the source and the detectors unless we know all the physical features of the particles and the intervening matter, energy and space. We do however, know the extremes: The particles can have the same spin, (both up, or both down) giving them a correlation of +1, or they can have opposite spin, giving a correlation of -1. Therefore, actual correlation values will be equal to one of, or lie somewhere between, these two extremes.

If the particles exist in a nonlocal reality, unknown physical features (sometimes called hidden variables) cannot effect the correlation of the particles before they arrive at the detectors, since quantum theory tells us that the particles do not exist until they register, and so their correlation depends solely upon the angle of orientation of the detectors. It is easily shown, using the simple concept of probability and high-school trigonometry, that the correlation is equal to the negative of the cosine of the angle of orientation between the detectors. See Figure A-1. (This functional relationship has also been confirmed by experiment.)

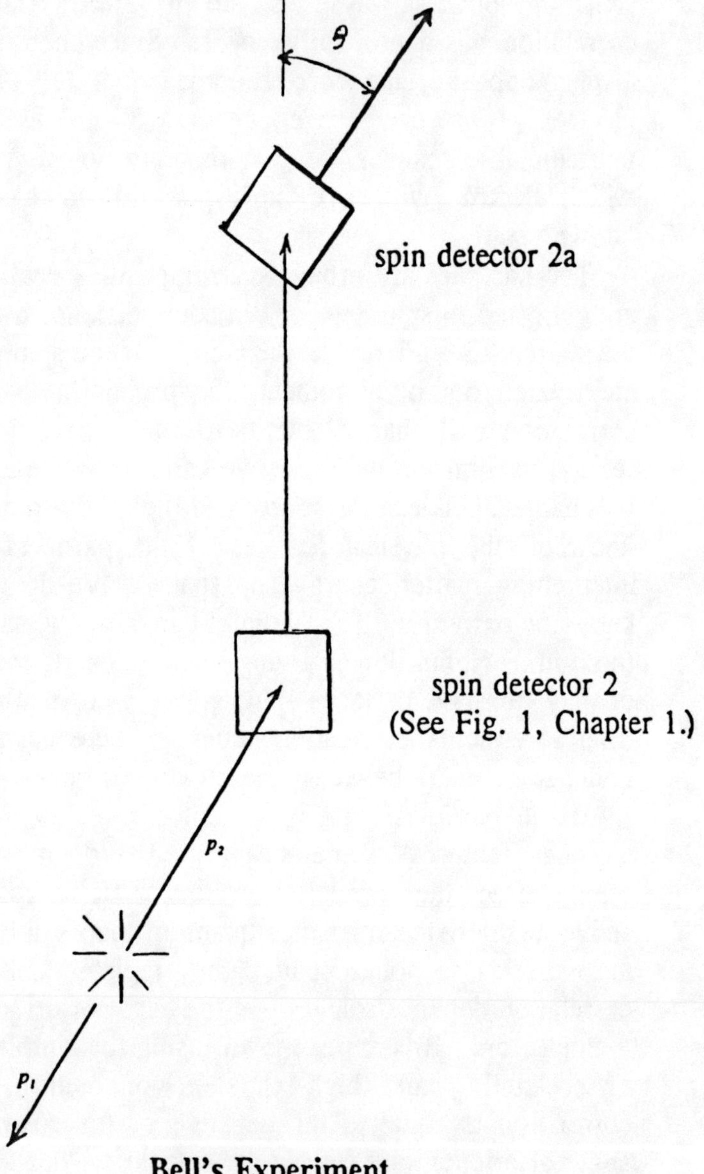

Bell's Experiment
Figure A-1

If the detectors are aligned perfectly, the angle is zero and the correlation is -1. For an angle of 30 degrees, the correlation is - 0.866, for 45 degrees, - 0.707, for 135 degrees, +0.707, and for 180 degrees, +1. Notice that, just like in the local-reality case, the values range from -1 to +1.

While the correlation for a specific pair of particles in a local reality might be different from their correlation in a nonlocal reality for the same angle of orientation, the exact values could not be calculated for the local reality case since the physical details of elementary particles are below the lower limit, or horizon, of even our instrumentally extended perception. Because the range of correlation in both cases was - 1 to + 1, it appeared that there was no way to distinguish between local and nonlocal reality.

The beauty of Bell's theorem is that he found a way to go beyond this apparent impasse. He derived a formula for calculating the correlation between particles arriving at the two distant detectors by making measurements, changing the angle of orientation of the detectors and making another set of measurements. Under this, more complicated arrangement, in a local reality the correlation between particles arriving at the detectors is equal to the sum of the joint probabilities for the two detectors before and after the angles of orientation are changed. The expression for this correlation and its maximum and minimum values can be derived as follows:

Let θ be the angle of orientation at A, ϕ the angle of orientation at B, $P(\theta)$ = the probability that a given spin orientation will be detected at A, and $P(\phi)$ = the corresponding probability at B. When the detectors are aligned, the spin of the particles detected at A and B will always be the same as their spin was at the source. Thus the spin detected at A will always be the opposite of that detected at B. In this case, $\theta = \phi = 0$, $P(\theta) = -1$, $P(\phi) = +1$, and the correlation between observations at the two detectors = $P(\theta,\phi) = -1$ times $+1 = -1$. If we let $P(\theta',\phi)$ represent the probability of correlated pairs with detector A rotated to a new angle, θ', with B still at the original angle, ϕ; $P(\theta,\phi')$ will represent the probability of correlated pairs with detector B rotated to a new angle, ϕ', with A unchanged; and $P(\theta',\phi')$ will represent the probability when both angles have been changed.

The summation of the joint probabilities for the two detectors before and after the angles of orientation are changed is equal to the expression:

$$P(\theta,\phi) - P(\theta,\phi') + P(\theta',\phi) + P(\theta',\phi') - P(\theta') - P(\phi)$$

As demonstrated above, for local reality the first term is equal to negative unity and the values for each of the others range from + 1 to - 1. With this knowledge, and the original correlation of the particle pair, we can demonstrate, by calculation of all the maximum and minimum values of each probability, that the total value of this expression is always ≤ 0. Also, if we substitute maximum and minimum values for $P(\theta')$ and $P(\phi)$, we see that:

$$-2 \leq P(\theta,\phi) - P(\theta,\phi') + P(\theta',\phi) + P(\theta',\phi') \leq +2$$

Considering the orientation of the detectors, the expression between the inequality signs represents the total expected value of the correlation in a local reality of the measurements made on pairs of particles in an experiment where the orientation angles of the detectors are changed during the experiment. While we have no way of calculating the exact values of the probabilities in this local-reality case, we know that the maximum value possible for the correlation in local reality, call it C_L, is +2, the minimum possible -2. That is,

$$-2 \leq C_L \leq +2$$

For nonlocal reality, while it is possible to calculate the probability at each detector for every value of θ and ϕ, quantum theory will not allow us to multiply individual probabilities to get the joint probabilities as we did in C_L. Instead, the probabilities are functions of the angle between the orientation of the two detectors. In the case where the detectors are rotated to an angle of 45 degrees after the first measurement, Bell calculated the correlation for the nonlocal reality predicted by quantum mechanics to be equal to + 2.83. In other words, if we perform this experiment in a world where elementary particles are objects in the same way that we experience baseballs or raindrops to be objects, C_L will always be less than + 2. On the other hand, in the type of world posited by the Copenhagen interpretation, with elementary particles that do not exist locally until they are captured in a way that can be measured and observed, in an experiment with the collectors rotated 45 degrees, the correlation will be equal to exactly 2.83. Thus, Bell reasoned that an actual experiment, performed with correlated pairs of electrons, photons, or other elementary particles, would yield either some value, C_L, less than +2, or the nonlocal value, C_{NL}, equal to 2.83. The results of such an experiment would reveal the real world of elementary particles as either local or nonlocal, and would finally determine who was right, Einstein and the determinists, or Bohr and the quantum physicists.

APPENDIX B

THE ASPECT EXPERIMENT

Bell's theorem, published in 1964, provided, at least in principal, a way to determine the actual nature of quantum reality. Was Einstein right? Is the physics of elementary particles consistent with the mechanics of larger particles that can be observed directly? Do electrons, photons, and other quanta behave like miniature baseballs, or is the Copenhagen interpretation of quantum mechanics correct? Bohr, Heisenberg and others held the view that elementary particles do not exist apart from the apparatus of observation and have no localized physical form until they are detected by some physical means that forces the collapse of the probabilistic wave function that describes the whole range of their possible states. This view is known as the Copenhagen interpretation of quantum mechanics.

Bell's theorem proved that, under certain conditions (see Appendix A) experimental results would be decisively different, depending upon whether elementary particles are local, deterministic phenomena or nonlocal, holistic events ultimately dependent upon observation. As such, it was a challenge to experimental physicists everywhere. Could they design and construct a real laboratory experimental arrangement which could produce initially correlated particle pairs and measure their final correlation upon detection at opposite ends of the laboratory with sufficient accuracy to prove which prediction was correct? If they could, the measurements would either result in a correlation less than + 2, implying classical local interactions, or the correlation would be greater than + 2, proving Bohr's Copenhagen interpretation correct, and the matter would finally be settled.

Accepting the challenge, a number of experimental physicists and engineers set about the task of designing the appropriate equipment. The efficiency of electron spin detectors proved to be a difficult problem. Error introduced by this problem could largely mask the results and so, it had to be solved before Bell's experiment could be performed. They determined that if they used photons instead of electrons, the problem of collector efficiency could be overcome. It was well known that certain atomic reactions produce photons pairs with 100% correlated polarization. And polarization is a physical feature that is easy to detect. After all, the Copenhagen interpretation had been proposed for all elementary particles, not just as an oddity associated with electrons. To avoid the EPR paradox, it had to apply to all elementary particles.

In 1972, John Clausner and Stuart Freedman, working at UC Berkeley, produced results that were fairly convincing, at least to some physicists. They used calcium atoms which, when heated, emit photon pairs in opposite directions with identical polarization. Polarization detectors, set up on opposite sides of the source, were oriented at 22.5 degrees and then at 67.5 degrees (a difference of 45 degrees). This experimental setup was, in general, similar to the setup described in Chapter 1 and Appendix A, only instead of spin detectors, we have polarization detectors.

The resulting correlation was greater than + 2. But the majority of physicists, trained in scientific materialism, were not convinced. Maybe the detectors, not the particles, were signaling each other. A subtle electronic impulse in the equipment linking the two detectors might be causing the unwelcome correlation. Surely, they reasoned, the excess correlation could be explained in some way that would not involve the heretical ideas of nonlocality and observer participation.

Experimental physicists, including Alain Aspect and colleagues in France devised new, improved experimental setups designed to rule out electronic feedback communications between the detectors. Their results were even more convincing than those that had gone before. In 1982, in an experimental setup for which quantum theory predicted a correlation of 2.70, the actual result obtained by Aspect and his colleagues was 2.697. Could there be <u>any</u> answer other than quantum connectedness? Well ... remember Einstein's refusal to accept quantum mechanics as anything more than a temporary statistical estimate of the real underlying details? What if photons have fine features allowing communication between the pair via some, as yet unknown, physical wave travelling at the speed of light? This could account for the excess correlation without violating current physical theory. Faster than light speed communication, let alone the <u>instantaneous</u> communication implied by the Copenhagen interpretation, violates the theory of relativity and opens physical theory to all sorts of possibilities that are unacceptable to scientific materialism. Most physicists thought that there <u>had</u> to be another answer..

Aspect ruled out light-speed communication between the photons as an explanation by changing the orientation of the detector at A long enough after the photons left the source that no signal from the left-hand photon could reach the right-hand photon before it arrived at detector B. Even with this precaution of assuring what became known as "Einstein separability", the results continued to be in excellent agreement with the quantum connectedness predictions.

With the clear-cut results of the Aspect experiment, we have to conclude that the existence of pairs of elementary particles formed from a common source, produced by a single event, can only be described in terms of a range of possible states between the source and their receptors. They do not produce the effects of separate physical entities until they impinge upon individual receptors, and these effects cannot be demonstrated to exist without conscious observation. These conclusions are consistent with the Copenhagen interpretation of quantum mechanics.

The experimental results of Aspect and others put believers in materialistic determinism in a difficult position. If the results are to be explained by hidden physical features within the quanta used in the experiment, the features must be such that the quanta are able to "sense" the arrangement of the experimental apparatus, and "know" what to expect at the detectors, in order to adjust their behavior to produce the exact correlative result that will fool the apparatus and the observer into thinking that quantum conectedness has occurred. Such extreme attempts to explain away the results in a way that will allow us to hang on to materialistic determinism force us to choose between a participatory, nonlocal universe and conscious, conniving quanta! What a dilemma for materialism!

APPENDIX C - PART I

A CALCULUS OF DISTINCTIONS PROOF
of
The Existence of Non-Quantum Receptors

Because physical phenomena are composed of quanta, the process of any observation is made up of and can be represented by a finite number of distinct elements. The most basic of all possible elements consists of a single quantum since, by the basic principle of quantum theory, a quantum cannot be divided. Therefore we will choose the quantum as the basic distinction of our calculus. See Appendix D for the definition of distinction.

An observation involves the origination, transmission, and reception of information in the form of quanta which we know as photons and electrons. The most basic descriptive system in the process of observation consists of three elements: a source, a quantum, and a receptor. Since all physical phenomena are composed of quanta, this description holds for each information transfer step of an observation, whether we are describing a photon from a light source being received by the retina of an eye, an electron being transferred from source to receptor across a neural synapse in the optic nerve, or a similar transfer in the brain. For the purpose of description and logical treatment in the Calculus of Distinctions, it is not necessary to define the size, location, or physical characteristics of the source, quantum, or receptor. We are dealing with them solely on the most basic level, i.e., as <u>distinctions</u> (see Appendix D). The basic process of the transfer of information in an observation may be described in the Calculus of Distinctions as follows:

A region of space, S_i, is divided into two distinct parts, **A** and **B**. By the definition of distinction, page 317, Appendix D, and logical interpretation, Table 1, page 76, <u>Infinite Continuity</u>, the state of S_i, resulting from the distinction of **A** from **B**, can be represented by:

I-1.) $$S_1 = \overline{\overline{A}|\ \overline{B}|}|\ ,$$

where **A** and **B** represent the contents or characteristics of the subregions of S_i that constitute the nature of their distinction. For the purposes of this discussion, let region **B** be the source of at least one quantum particle, q_1, then,

I-2.) $$S_1 = \overline{\overline{A}|\ \overline{\overline{B_1}|\ \overline{q_1}|}|}|$$ where $\mathbf{B} = \overline{\overline{B_1}|\ \overline{q_1}|}|$

and $\mathbf{B_1}$ = all contents of **B** except q_1, if there are any.

After some period of time, the quantum q_1 impacts on the contents of region **A**, imparting the information content it carries to a quantum, q_2, in **A**. So that

I-3.) $$S_2 = \overline{\overline{\overline{A}|\ \overline{q_2}|}|\ \overline{B_1}|}|$$ now represents the state of S_i. It is important to note that we do not have to concern ourselves with the questions of how the information gets from region **B** to region **A**, or how much time elapses between states S_1 and S_2. The Copenhagen interpretation (See Chapter 2.) tells us that the quantum particle has no path or local physical existence until it impacts on the receptor **A**. Consequently, we only need to concern ourselves with the state or contents of the distinguished regions of the space **S**. Similarly, we do not have to concern ourselves with the questions of the mechanism of generation of q_1 or q_2, or the exact description of the information carried by them. We know, for example, that photons and electrons perform these functions in the processes involved in making an observation, and that information is transferred from the object to the observer. While there may be millions or billions of quanta involved in each step of an actual observation, it is sufficient for our purposes here to consider the only the distinct elements in the chain of events transferring the information of one quantum from object to observer.

In an observation there are a number of transfer links like the transfer from **B** to **A** described above. For instance, the first transfer may be that of a photon from a light source to the object of observation, the second, a photon from the object to the eye of an observer or to a photoelectric plate, the third may be an electron in the current generated by the photoelectric plate to an atom in a conductor initiating an electric current in a wire, etc., eventually to the transfer of an electron in a neural synapse in the brain of the observer. We may subdivide the region S_i further to describe the various physical systems involved in the transfer of the information of $\mathbf{q}_1$ from the first source involved in the observation to the first receptor, which by the impact of $\mathbf{q}_1$ is caused to act as a source for $\mathbf{q}_2$, which then registers on a second receptor, causing it to act as a source for $\mathbf{q}_3$, and so on. The number of subdivisions of S_i is finite and depends on the number of quanta involved in the actual chain of information transfer in the observation being described. To reflect the chain of quantum transfers in an observation, we may write S_1 as follows:

I-4.) $\mathbf{S}_1 = \overline{\mathbf{A}\rceil\overline{\mathbf{B}}\rceil}\rceil = \overline{\mathbf{C}\rceil\overline{\mathbf{B}_1\rceil\overline{\mathbf{q}_1}\rceil}\rceil}\rceil$, where $\mathbf{B}_1$ $\mathbf{q}_1$ represents the elements of quantum transfer in **A** and **B** except for **C**, the consciousness of the observer. In addition, we may subdivide $\mathbf{B}_1$ into **n** subdivisions $\mathbf{b}_n$, $\mathbf{b}_{n-1}$, $\mathbf{b}_{n-2}$, ... $\mathbf{b}_3$, $\mathbf{b}_2$, $\mathbf{b}_1$, where **n** is the total number of quantum transfers in the chain of information transfers. Then expression I-4 becomes:

I-5.) $$S_1 = \overline{C\rceil\; b_n\rceil\; b_{n-1}\rceil\; b_{n-2}\rceil \;\ldots\; b_3\rceil\; b_2\rceil\; b_1\rceil\; q_1\rceil}$$

and the sequence

I-6.) $$S_2 = \overline{C\rceil\; b_n\rceil\; b_{n-1}\rceil\; b_{n-2}\rceil \;\ldots\; b_3\rceil\; b_2\rceil\; q_2\rceil\; b_1\rceil}$$

I-7) $$S_3 = \overline{C\rceil\; b_n\rceil\; b_{n-1}\rceil\; b_{n-2}\rceil \;\ldots\; b_3\rceil\; q_3\rceil\; b_2\rceil\; b_1\rceil}$$

and so forth, represents the successive transfer of information in the form of quanta, until:

I-8.) $$S_1 = \overline{C\rceil\; b_n\rceil\; q_n\rceil\; b_{n-1}\rceil\; b_{n-2}\rceil \;\ldots\; b_3\rceil\; b_2\rceil\; b_1\rceil}$$

The states of S_i represented by I-5 through I-7 are depicted by the diagrams in Figure C-1. Each distinction, A, B, C, ... acts first as receptor, then as source in the chain of information transfers.

Up to this point, all the transfers have been simple quantum transfers involving the relay of the specific information carried by the sequence of quanta, q_1, q_2, q_3, ... q_n, and the requirements of the Copenhagen interpretation have been met in each case, with each quantum being precipitated from the schroedinger wave of probabilities by impacting on a physical receptor. The final receptor, C, has to be part of the consciousness of the observer. But what is the nature of this receptor? If it is composed of a specialized group of cells in the brain, or any other part of the observer's body, then it is composed of quanta of matter and energy, like every other source/receptor in the chain. But the moment we identify

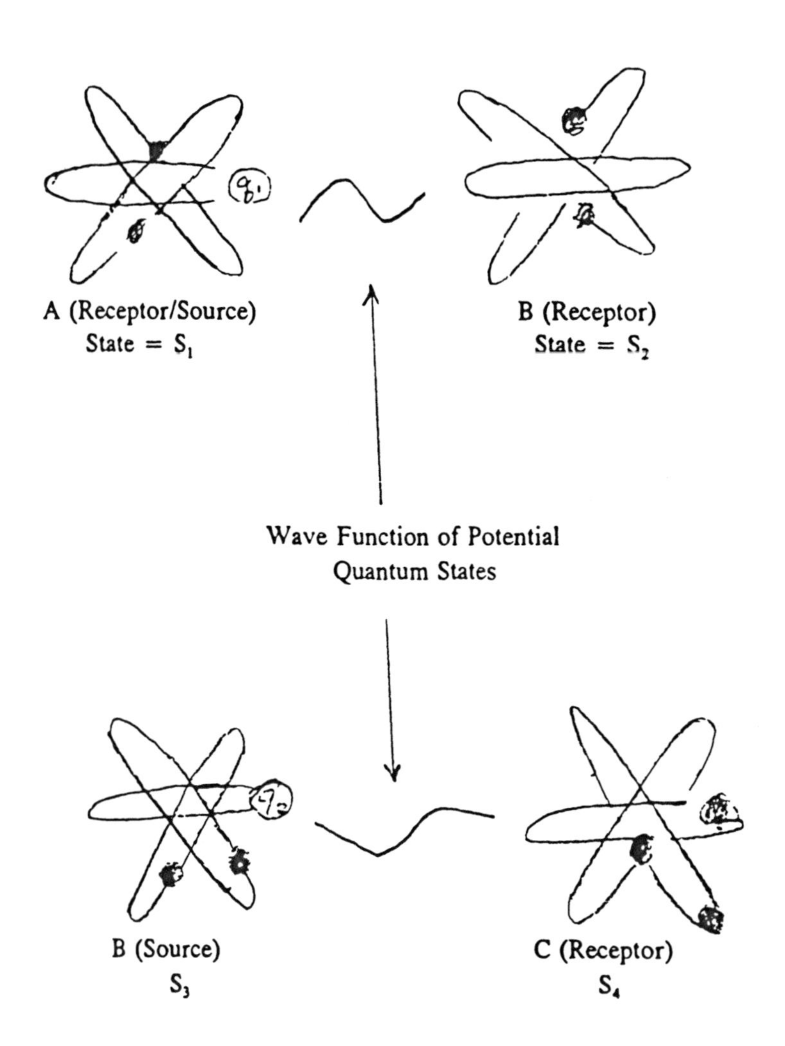

Sources and Receptors
Figure C-1

this receptor in the brain, it becomes an object which can be observed by the observer and therefore is not the observer. Something else has to receive the information of q_n, to integrate it with the billions of other quanta of information, and to form an image from them. This something else has to function as the final receptor, and therefore, another step is added to the chain of transfers. To express this, we divide $\mathbf{C}$ into $\mathbf{C}_1$ and $\mathbf{b}_{n+1}$ and we have:

$$\text{I-9.)}\ S_{n+1} = \mathbf{C}_1 \rceil\, \mathbf{b}_{n+1} \rceil\, q_{n+1} \rceil\, \mathbf{b}_n \rceil\, \mathbf{b}_{n-1} \rceil\, \mathbf{b}_{n-2} \rceil \ldots\ \mathbf{b}_3 \rceil\, \mathbf{b}_2 \rceil\, \mathbf{b}_1 \rceil$$

And now we must ask: What is the nature of $\mathbf{C}_1$, and how does it receive the information from q_{n+1}? The Copenhagen interpretation established the fact that each quantum requires a receptor to become a local phenomenon, and if we insist that each receptor must be physical, i.e., made up of quanta of matter and energy, then each additional transfer requires that we move another sub-partition of $\mathbf{C}$ into the sequence of physical source/receptors, and so on, ad infinitum. But this is prohibited by the basic principle of quantum theory: Planck's constant determines the size of a quantum, and it cannot be subdivided. This places a bottom on our descent and we have, therefore, reached a contradiction by infinite descent.

It is clear that the chain must stop with some finite number of receptors., and the final one, at least, must possess or be a part of the consciousness of the observer. If this final receptor is composed of physical quanta, we have some serious problems. We simply cannot explain consciousness in terms of quanta. If the information relayed by the chain of quanta is not passed on to another receptor, and remains as an effect on the last physical receptor, how are the many images of the object of observation formed, and how are all the other functions of consciousness performed? How can the entire panorama on a starry night, or a view from a mountain top be re-created moment-by-moment as the photons enter our eyes? How can all the information of years of observations, plus the billions of bits of information from the other senses, all received at the rate of millions of bits per second, possibly be sorted, integrated, stored, and recalled by a limited collection of brain cells made up of quanta that cannot be divided? How can physical structure, however complex, ever give rise to consciousness?

If, on the other hand, we define C as a non-quantum receptor at the end of the chain of quantum sources and receptors, and as a necessary part of the limited, nonlocal consciousness of the observer, then the problems inherent in the physical limitations of quanta as the media of information storage, image formation, and the psychological functions of consciousness are resolved. The brain contains the necessary complexity of structure to function as a filter, reducing and organizing quantified information for input into consciousness where continuous images may be formed and stored like holograms within the nonlocal space of consciousness. It is likely that the brain also serves the consciousness of the observer as an output data processor and temporary storage unit.

The nonlocality of consciousness suggests that it is effectively infinitely continuous wherever it exists, filling all the space occupied by the physical quanta of matter and energy, in global contact with the quanta making up specialized structures of the brain. Without the "bottom" of physical quanta, and with the feature of nonlocality, the space of consciousness may be divided indefinitely. With this understanding, we can write a new expression for the transfer from S_n to S_{n+1} :

I-10.) $$S_{n+1} = \overline{C \rceil \overline{q_{n+1}} \rceil \rceil \overline{b_n} \rceil \overline{b_{n-1}} \rceil \overline{b_{n-2}} \rceil \ldots \overline{b_3} \rceil \overline{b_2} \rceil \overline{b_1} \rceil \rceil}$$

where q_{n+1} is the information content relayed from the object of observation by the sequence of quanta, q_1, q_2, q_3, ... q_n, and C is non-physical, i.e., not composed of quanta of matter and/or energy. Thus the contradictions of the infinite descent of receptors and the mathematical impossibility of the representation of objective reality and the many complex functions of consciousness by the quanta making up the brain are avoided. This completes the proof of the existence of non-quantum receptors.

APPENDIX C - PART II

A CALCULUS OF DISTINCTIONS PROOF

of

The Existence of a Non-Quantum Receptor Prior to the Appearance of the First Quantum

We begin this proof by writing a Calculus of Distinctions expression describing the current state of the universe. Let U_n represent the current state of the universe. If it is true that the universe is expanding, and that the quantum principle and the general theory of relativity are valid, then U_n is composed of a finite (although <u>very</u> large) number of elementary distinctions, and may be fully described by the following Calculus of Distinctions expressions:

II-1.) $$U_n = \overline{S_n\rceil\ \overline{S_{n-1}}\rceil\ \overline{S_{n-2}}\rceil \cdots \overline{S_3}\rceil\ \overline{S_2}\rceil\ \overline{S_1}\rceil\rceil},$$

where **n** is the number of distinct physical systems, S_i, in U_n, indexed from right to left in the order in which they were formed in the expanding universe, and

II-2.) $$S_i = \overline{R_{i,j}\rceil B_{i,j}\rceil q_{i,j}\rceil\ R_{i,j-1}\rceil B_{i,j-1}\rceil q_{i,j-1}\rceil \cdots R_{i,3}\rceil B_{i,3}\rceil q_{i,3}\rceil\ R_{i,2}\rceil B_{i,2}\rceil q_{i,2}\rceil\ R_{i,1}\rceil B_{i,1}\rceil q_{i,1}\rceil},$$

where **j** is the number of quantum source/receptor systems in S_i, and the **j**'s for the various S_i's are positive integers ranging from 1 to very large, but finite numbers. **R**, **B**, and **q** represent the receptor, source, and quantum, respectively in a set of quantum information transfer elements as described in Part I. The individual quantum transfer systems are of the form:

II-3.) $$\overline{R_{i,k}\rceil B_{i,k}\rceil q_{i,k}\rceil\rceil} = \overline{C_{i,k}\rceil\ b_{i,k,m}\rceil\ b_{i,k,m-1}\rceil\ b_{i,k,m-2}\rceil \cdots}$$

$\overline{\overline{\overline{b_{i,k,p}}\rceil\, q_{i,k,p}}\rceil\rceil \cdots \overline{b_{i,k,2}}\rceil\, \overline{b_{i,k,1}}\rceil\rceil}$, where $\mathbf{m}$ is the number of source/receptors in $\mathbf{B}_{i,k}$, $\mathbf{k}$ ranges from 1 to $\mathbf{j}$, $\mathbf{C}_{i,k}$ is the final receptor in each transfer system and is a conscious, non-quantum receptor (as in Part I) in at least some of the systems, and the quantum $q_{i,k,p}$ may be anywhere in the chain of $\mathbf{b}_{i,k,1}$ to $\mathbf{b}_{i,k,m}$ source/receptors.

To complete the description of $\mathbf{U}_n$, we note that, in the current universe, most of the source/receptors are composed of large numbers of quanta of matter and energy, and so:

II-4.) $\overline{b_{i,k,r}}\rceil = \overline{\overline{q_{i,k,s}}\rceil\, \overline{q_{i,k,s-1}}\rceil\, \overline{q_{i,k,s-2}}\rceil \cdots \overline{q_{i,k,3}}\rceil\, \overline{q_{i,k,2}}\rceil\, \overline{q_{i,k,1}}\rceil\rceil}$,
where $\mathbf{r}$ ranges from 1 to $\mathbf{m}$ and $\mathbf{s}$ is the number of quanta in $b_{i,k,r}$ and may range from 1 to a very large, but finite number.

Since the state variables in relativistic and quantum equations are commonly expressed as vectors in Minkowski space or tensors in Hilbert space, one may naturally tend to think of the $\mathbf{S}$, $\mathbf{C}$, $\mathbf{B}$, and q variables in these Calculus of Distinctions expressions as vectors. Therefore, it must be emphasized here that they should not be thought of as vectors in Hilbert space, since there are no numerical values, quantities, or magnitudes attached to any of these variables. The expression S_i simply signifies the distinction of one subregion of a universe from the rest of that universe, as defined in Appendix D. The indexing of a symbol marking or distinguishing a space or subregion with a counter such as $\mathbf{i}$, $\mathbf{j}$, $\mathbf{k}$, $\mathbf{l}$, $\mathbf{m}$, $\mathbf{n}$, etc., simply indicates that the symbol is a variable in the sense that a number of such distinctions may exist, and does not imply that the symbol represents any particular magnitude, direction, or size. It should be born in mind that the logic of the Calculus of Distinctions applies prior to the consideration of the assignment of magnitudes to variables in the process of the development of mathematical tools.

Returning to the expressions II-1, II-2, II-3, and II-4, defining the current state of the universe, if we trace the sequence of events backward from the current state, U_n , toward the mathematical singularity of the big bang, members of the series U_n , U_{n-1} , U_{n-2} , ... become increasingly less complex as the individual quanta composing the physical form and structure of the universe are resurrected from their respective receptors and returned to their sources. A series of expressions describing this process may be derived as follows: First, substitute the expressions defined by II-4 into II-3; second, substitute the resulting series into II-2; and then substitute the result of that into II-1 to obtain a complete, detailed expression for U_n. Next, the expression for U_{n-1} is obtained when all the quanta, $q_{n,k,1}$ through $q_{n,k,s}$, are reabsorbed into their sources in $B_{n,k,r}$, and the space S_n is void of physical objects, and therefore, no longer distinguished from S_{n-1}. And so:

II-5.) $$U_{n-1} = \overline{S_{n-1} \overline{S_{n-2} \overline{S_{n-3} \ldots \overline{S_3 \overline{S_2 \overline{S_1}}}}}},$$

which may be expanded by the substitutions described above, as follows:

$$U_{n-1} = \overline{\overline{R_{n-1,j} \overline{B_{n-1,j} \overline{q_{n-1,j}}}} \; \overline{R_{n-1,j-1} \overline{B_{n-1,j-1} \overline{q_{n-1,j-1}}}} \ldots}$$

$$\overline{\overline{R_{n-1,1} \overline{B_{n-1,1} \overline{q_{n-1,1}}}} \; \overline{R_{n-2,k} \overline{B_{n-2,k} \overline{q_{n-2,k}}}} \; \overline{R_{n-2,k-1} \overline{B_{n-2,k-1} \overline{q_{n-2,k-1}}}} \ldots}$$

$$\ldots \overline{R_{n-2,1} \overline{B_{n-2,1} \overline{q_{n-2,1}}}} \ldots \overline{R_{1,1} \overline{B_{1,1} \overline{q_{1,1}}}}$$

where j is the number of systems in S_{n-1} , k is the number of systems in S_{n-2} , and so forth. Then :

$$\overline{R_{n-1,j}\overline{B_{n-1,j}\overline{q_{n-1,j}}}} = \overline{C_{n-1,j}\overline{b_{n-1,j,m}\overline{b_{n-1,j,m-1}\cdots\overline{b_{n-1,j,p}\overline{q_{n-1,j,p}}}}}}$$

$$\cdots\overline{b_{n-1,j,2}\overline{b_{n-1,j,1}}},$$

$$\overline{R_{n-2,k}\overline{B_{n-2,k}\overline{q_{n-2,k}}}} = \overline{C_{n-2,k}\overline{b_{n-2,k,u}\overline{b_{n-2,k,u-1}\cdots\overline{b_{n-2,k,v}\overline{q_{n-2,k,v}}}}}}$$

$$\cdots\overline{b_{n-2,k,2}\overline{b_{n-2,k,1}}}, \ldots$$

$$\overline{R_{1,1}\overline{B_{1,1}\overline{q_{1,1}}}} = \overline{C_{1,1}\overline{b_{1,1,1}\overline{q_{1,1,1}}}} \text{ and}$$

$\overline{b_{x,y,m}} = \overline{q_{x,y,s}\overline{q_{x,y,s-1}\overline{q_{x,y,s-2}\cdots\overline{q_{x,y,3}\overline{q_{x,y,2}\overline{q_{x,y,1}}}}}}}$, where **x** ranges from **1** to **n-1** in S_{n-1}, from **1** to **n-2** in S_{n-2}, etc.; **y** ranges from **1** to **j**, **k**, etc., **m** is the number of **b**'s in each **B**, and **s** is the number of distinctions, **q**, in each **b**.

Clearly, the fully expanded Calculus of Distinctions representation of the regressive sequence U_n, U_{n-1}, U_{n-2}, ... U_3, U_2, U_1, U_0 would require an enormous amount of space. For example, using the minimum number of expressions needed to define the sequences in U_n alone, would produce 1,440 distinctions and require a system of indices that would quickly exhaust both the English and Greek alphabets. Fortunately, we can conserve space by representing U_n with the greatly abbreviated form:

$$U_n = \overline{C_{n,k}\overline{b_{n,k,m}\overline{b_{n,k,m-1}\cdots\overline{b_{n,k,p}\overline{q_{n,k,p}}}\cdots\overline{b_{n,k,2}\overline{b_{n,k,1}}}}}}$$

$$\overline{C_{n,k-1}\overline{b_{n,k-1,d}\cdots\overline{b_{n,k,p}\overline{q_{n,k,p}}}\cdots\overline{b_{n,k-1,2}\overline{b_{n,k-1,1}}}\overline{C_{n,k-2}\overline{b_{n,k-2,e}\cdots}}}}$$

$$\overline{b_{n,k-2,2}\overline{b_{n,k-2,1}}}\cdots\overline{C_{n-1,f}\overline{b_{n-1,f,g}\cdots}}\cdots\overline{C_{1,1}\overline{q_{1,1,1}\overline{b_{1,1,1}}}}$$

And if we represent the sequences $\overline{C_{n,k}\overline{b_{n,k,m}\overline{b_{n,k,m-1}}}}\cdots$

$\overline{b_{n,k,2}\overline{b_{n,k,1}}}\ \overline{C_{n,k-1}\overline{b_{n,k-1,d}\cdots}}$ with $\overline{C_n\overline{B_n\overline{q_n}}}$, and

$$C_{n-1,f}\; b_{n-1,f,g} \cdots C_{n-1,f-1}\; b_{n-1,f-1,q} \cdots \quad \text{with} \quad C_{n-1}\; B_{n-1}\; q_{n-1},$$

and so forth, we can further abbreviate U_n and the regressive sequence to:

$$\text{II-6.)}\quad U_n = C_n\; B_n\; q_n\; C_{n-1}\; B_{n-1}\; q_{n-1} \cdots C_{n-2}\; B_{n-2}\; q_{n-2} \cdots C_1\; b_1\; q_1$$

$$U_{n-1} = C_{n-1}\; B_{n-1}\; q_{n-1}\; C_{n-2}\; B_{n-2}\; q_{n-2} \cdots C_1\; b_1\; q_1$$

$$U_{n-2} = C_{n-2}\; B_{n-2}\; q_{n-2}\; C_{n-3}\; B_{n-3}\; q_{n-3} \cdots C_1\; b_1\; q_1$$

•   •   •

•   •   •

(As we trace the sequence backward and each S_i subregion collapses in on itself, the corresponding sequence of distinctions representing the quanta/source/receptor systems for that subregion disappear from the expression representing the state of the universe. For the last few U_i expressions in the sequence, it is possible to write more of the expression within a reasonable space. Therefore, we can now write them in a less abbreviated form.)

•   •   •

•   •   •

$$U_3 = C_{3,a}\; b_{3,a,b}\; b_{3,a,b-1} \cdots b_{3,a,c}\; q_{3,a,d} \cdots b_{3,a,2}\; b_{3,a,1}$$

$$C_{3,a-1} \cdots C_{2,e}\; b_{2,e,f}\; b_{2,e,f-1} \cdots b_{2,e,g}\; q_{2,e,g} \cdots b_{2,e,2}\; b_{2,e,1}\; C_{2,e-1} \cdots$$

$$C_{1,h}\; b_{1,h,i}\; b_{2,h,i-1} \cdots b_{2,h,j}\; q_{2,h,j} \cdots b_{1,h,2}\; b_{1,h,1}\; C_{1,h-1} \cdots$$

$$U_2 = C_{2,c}\, b_{2,c,f}\, b_{2,c,f-1} \cdots b_{2,c,r}\, q_{2,c,r} \cdots b_{2,c,2}\, b_{2,c,1} \quad C_{2,c-1} \cdots$$

$$C_{1,h}\, b_{1,h,i}\, b_{1,h,i-1} \cdots b_{1,h,s}\, q_{1,h,s} \cdots b_{1,h,2}\, b_{1,h,1} \quad C_{1,h-1} \cdots$$

$$U_1 = C_{1,h}\, b_{1,h,i}\, b_{1,h,i-1} \cdots b_{1,h,j}\, q_{1,h,j} \cdots b_{1,h,2}\, b_{1,h,1} \quad C_{1,h-1}$$

$$b_{1,h-1,k}\, b_{1,h-1,k-1} \cdots b_{1,h-1,m}\, q_{1,h-1,m} \cdots b_{1,h-1,2}\, b_{1,h-1,1} \quad C_{1,h-2} \cdots$$

$$C_{1,1}\, q_{1,1,1}\, b_{1,1,1} .$$

And as the physical systems of the last S_i collapse in order, we eventually have only the first receptor, source, quantum set that formed in the expanding universe left:

$$C_{1,1}\, q_{1.1,1} \quad b_{1,1,1} .$$

After expression II-3, we noted that the final receptors, $C_{i,k}$, are conscious receptors in some cases. The author and reader are ready evidence of this fact. As we look at the expanding universe in reverse, two very important and closely related questions must be asked: First, what happens when the universe is regressed to the point where no organic life forms exist to function as conscious observers? And second, what was the nature of the first receptor/quantum /source? These may sound like difficult questions at first, but the Copenhagen interpretation, Bell's theorem and the results of the Aspect experiment give us the means to answer them.

In order to facilitate our discussions, we will use the acronym CBA to denote the following. C: the Copenhagen conclusion that no phenomena exist until they register on a receptor. That is to say, elementary particles or waves do not exist as localized physical phenomena travelling through space. Their only physical existence separate from source, receptor and apparatus occurs when they impact upon a receptor in a physically measurable way. **B**: Bell's theorem demonstrates mathematically that the elementary particles of physical reality can be either separate, local phenomena, or wholistic, nonlocal phenomena, but not both. **A**: the results of the Aspect experiment, the most widely accepted of a number of similar experiments, proved that reality is nonlocal. The results of the Aspect experiment, along with the delayed-choice two-slit experiment and others, confirms the Copenhagen interpretation.

Concerning the first question: In our regression of the quantum processes involved in the expanding universe of matter and energy, we must eventually reach the stage where all the quanta composing the last (first in the expanding universe) physical forms, complex enough to perform conscious observations, have returned to their sources. At this point there are no localized non-quantum receptors existing within physical structures available to act as final receptors (see Part I). Beyond this point, in the direction of the big bang singularity, there will be no physical structure or order organized by living creatures. This would include structures for the purposes of self preservation (traps to capture prey, agriculture, etc.) or species propagation (nests, houses, etc.). It may be argued that life-supporting structures are insignificant relative to the larger structure of the universe, unless we consider the possibility that all structure that makes life possible, such as the structure of the hydrogen atom, the earth's peculiar balance of elements, our distance from the sun, etc., were brought about by consciousness at a deeper, more spatially extensive, nonlocal level.

If we assume that consciousness has no involvement prior to the appearance of some sort of life form, moving on back toward the beginning, we encounter a problem. CBA tells us that no elementary physical phenomena exist until they register on a receptor. If no quantum receptors exist, upon what do the first quanta register? The existence of <u>any</u> physical receptor at this stage raises the same question concerning its origin and we have the contradiction of infinite descent. Even if we posit a non-quantum physical state in the early stages of the universe due to the density of the substance of this early "proto-matter", at the point of the appearance of the first quantum, we still have the same problem.

In order to solve the problem of first cause, the first receptor had to be non-quantum and nonlocal. Since physical solutions to the contradiction of infinite descent are ruled out by the basic principle of quantum mechanics, only one possible solution remains. Because of CBA and the logic of infinite descent, we know that consciousness is a nonlocal, non-quantum substance capable of acting as a final receptor in observations. Furthermore, consciousness is the only *known* non-quantum substance with the properties of nonlocality and wholeness. Conclusion:

THE FIRST RECEPTOR <u>HAD TO BE</u> SOME FORM OF CONSCIOUSNESS

Returning to our Calculus of Distinctions regression, recall that just before the universe disappears into a mathematical singularity, it is simply:

$$U_1 = \overline{\overline{\overline{C_{1,1}}\, q_{1,1,1}}\, b_{1,1,1}}.$$

And in the final step of this regression, the receptor $C_{1,1}$ gives up the quantum $q_{1,1,1}$, which is re-absorbed into the source $b_{1,1,1}$, and so,

$$U_0 = \overline{\overline{C_0}\,\overline{b_0}} \,,$$

where b_0 is the unmanifested spectrum of all possible states of q_0, and C_0 is the original, or primary, nonlocal receptor. Beyond this point, no physical distinction has been made, and we cannot place distinction brackets over C_0 and/or b_0 and the logic of the Calculus of Distinctions (Table 1, Appendix D) implies that C_0 and b_0 are one and the same thing. Since C_0 is nonlocal, and b_0 is unmanifested possibilities, they are, at this point, not separate, nor logically separable.

APPENDIX D

THE CALCULUS OF DISTINCTIONS

INTRODUCTION

A most basic feature of consciousness is the ability to draw distinctions. No awareness, perception, knowledge, or description is possible without the recognition and/or formation of distinctions. Any conscious being first makes the distinction of self from other and then draws many distinctions of various types in the part of reality considered to be "other". All the different forms of information that make up the basis of awareness have one thing in common, they are made up of distinctions: distinctions of color, shape, texture, sound, taste, smell, etc. While these distinctions seem to be very different in nature, the conscious processing of information is the same for any distinction. It is this sameness that we must focus on, and in order to understand consciousness, we must develop a clear, precise definition of this most basic conscious act, the drawing of a distinction.

It is the thesis of this presentation that consciousness is not an abstraction, but a physical reality. It is a physical reality more subtle than the other forms of physical reality which are composed of quanta of matter and energy, that have historically been the subject of the science of physics. Just as we have developed a mathematical theory of matter and energy by studying the structure and interactions of matter and energy, we must attempt to discover the structure and nature of consciousness by studying its logical structure and the way it interacts with matter and energy through the drawing of distinctions.

The content of a distinction is anything which allows a conscious entity to distinguish one thing from another. To have substance or reality, the object of a distinction must have extent and content. Extent is usually defined in terms of time and space, and content is the feature which distinguishes the region in which the distinction extends from the rest of the universe. In order to make this concept useful, we must define the drawing of a distinction in a way that applies equally well to all types of distinctions. We may do this by noting that every distinction involves the recognition of three things: the existence of a distinguishing feature, the absence of that feature, and the transition from a state where the feature is present to a state where the feature is absent. For convenience in developing a mathematical notation, we may think of these three things as the marked state, the unmarked state, and the boundary between them.

While the choice of notation is somewhat arbitrary, until such a time as holographic projection becomes commonplace, we are limited to some two-dimensional form amenable to sheets of paper and the pages of books. For simplicity of expression, I have chosen to use a slight variation of the notation used by George Spencer Brown in his book <u>Laws of Form</u>. The drawing of a distinction will be indicated by the symbol $\urcorner$. The curved line may be thought of as the boundary, separating that which is distinguished from that which is not. A symbol or symbols under the distinction sign, such as $\overline{A}|$ or $\overline{B}|$ mark the content of the distinction. A symbol or symbols outside the distinction sign, such as $A\urcorner$ or $\urcorner B$, apply to the content of regions beyond the extent of the distinction.

For the purpose of describing the logical structure of consciousness through the drawing of distinctions, the symbols of the Calculus of Distinctions shall denote <u>only</u>

distinction, not quantity or magnitude. The conscious act of drawing a distinction is logically prior to the assignment of numerical value. While enumeration and counting are also distinctions, if they are introduced prematurely, as they are in conventional mathematics, the structure and form that flow from the pure act of the drawing of distinctions are masked and all but lost. If we wish to study the interaction of consciousness with matter and energy, we must start at the beginning, and the beginning is the drawing of a distinction.

Until a distinction is drawn, the state characteristic of consciousness is _void_. Since voidness is conceptually prior to enumeration and is, therefore, different from the value zero, we need a way to indicate this state. In many Calculus of Distinctions transformations the void state will need no symbolic representation other than a blank space, since voidness is the lack of any distinction whatsoever, but in situations where the lack of distinction must be indicated to avoid confusion, the symbol ¬ will be used.

A similarity between the Calculus of Distinctions and Boolean algebra will be recognized by some readers, but those readers should not let this similarity mislead them. Like all the other forms of mathematical and logical reasoning in general use today, before George Spencer Brown's _Laws of Form_, Boolean algebra was never connected with its roots in the conscious drawing of distinctions. The Calculus of Distinctions briefly derived and described in this Appendix reveals the ground in which all mathematics and logic rest. This ground is the logical structure of consciousness. The Calculus of Distinctions surpasses the scope of Boolean algebra in that it establishes the non-numerical arithmetic in which all forms of math and logic are based. It includes the root concepts that give rise to imaginary and complex numbers, which Boolean algebra does not do. Because of this, and other features related to logical structure, the Calculus of Distinctions shows us how Boolean algebra and symbolic logic fit into the broader framework of mathematical logic.

DEFINITIONS

Definition 1: CALCULATION

Calculation is the logical transformation of one form or expression into another equivalent but different form.

Example: $1 + 1 = 2$. The expression $1 + 1$ is transformed into the equivalent, but different form, 2, by the fundamental operation of addition.
Note that the act of drawing a distinction qualifies as a fundamental operation, since it transforms a void space into a different, but equivalent form, consisting of two parts: the region distinguished and the rest of the universe. This fundamental operation is even more fundamental than addition, subtraction, multiplication, and division, because it is a logical operation that must occur prior to them.

Definition 2: CALCULUS

A logical system developed for the purpose of calculation.

For the purpose of calculation, the state of a region may be called its value, and there are only two values defined at this point: distinct and void. A region differentiated by a distinction sign, ⏋, includes every distinguishing feature within that region. Distinctions within the region may be described by iterations of the symbol $\urcorner$ under the distinction sign applying to the whole region: $\overline{\urcorner\urcorner\urcorner}|$. Regions with the same value are equivalent and expressions describing the equivalent distinctions or combinations of distinctions of such regions will be linked by an equals sign. For example,

$\overline{A}| = \overline{B}|$ will indicate that the expressions on either side of the equals sign are reducible to the same value, either

$\urcorner$ = the state of being distinct, or $\not\urcorner$ = the void state.

Definition 3: PRIMARY EXPRESSIONS

A primary expression is an equivalence derived from the primary features of consciousness.

Consciousness exhibits nonlocality and wholeness, and it is also capable of focusing on discrete features within that wholeness. The nonlocality of consciousness is described by the equation:

$$\urcorner\urcorner = \urcorner \qquad \text{Ex. 1.}$$

This equation expresses the fact that the value of an expression does not depend upon the number, size, or location of the distinctions contained within the region distinguished by the expression. Brown calls this "the form of condensation".

The ability of consciousness to recognize discrete features is symbolized by the equation:

$$\overline{\overline{}|}\,| = \not\rceil \qquad \text{Ex. 2.}$$

which expresses the recognition of boundary. Consciousness becomes aware of a distinction by isolating the distinguishing feature or features from the rest of reality by moving its focus from another region, or from the void, across the boundary into the distinguished region. If the boundary is crossed again, as symbolized by Ex. 2, the focus has returned to the state outside the boundary, in this case the void state. Brown refers to this equation as "the form of cancellation".

Definition 4: SIMPLIFICATION

The value of an expression is equivalent to the value to which it can be simplified by applications of the primary expressions.

For example, given the expression

$$x = \overline{\rceil\,\rceil}|\ \rceil$$

x may be simplified as follows:

$$x = \overline{\rceil\,\rceil}|\ \rceil \quad = \overline{\rceil}|\ \rceil \qquad \text{by Ex. 1.}$$

$$\text{and } \overline{\rceil}|\ \rceil \quad = \rceil \qquad \text{by Ex 2.}$$

$$\text{Therefore,} \quad x = \rceil$$

Here we already begin to see how the Calculus of Distinctions may be used to develop simple logical structure without concern for magnitude or quantity. Why should we want to be able to do this? A great deal is being said and written these days about the similarity of the brain to a computer. This analogy can only be carried so far. While numerical calculations were first carried out mentally, the scope of activities carried out in the brain by consciousness far exceeds the manipulation of quantified information. Scientists in general, and mathematicians and physicists in particular, tend to forget that features that may be quantified, while not insignificant by any means, are abstracted from a much broader and richer reality because they are the simplest and easiest aspects of reality to study. To describe the structure of consciousness, we need a mathematical language not limited to the subset of realities that may be quantified.

QUANTUM DISTINCTIONS

In Chapter 2 we chose the quantum as the most basic distinction in our proof of the necessity of non-quantum receptors. If quanta are the basic distinctions underlying all of physical reality, and if quanta do not exist as local phenomena until received by non-quantum receptors -which we have identified with consciousness- then why aren't we directly aware of individual quanta such as electrons and photons? The answer is that we don't *sense* the effect of an individual quantum because of its size relative to the size of the atoms and molecules in the structure of our physical sensing organs. But, at some level, we probably are aware of individual quanta. We are, of course, aware of their effects in the aggregate. We can feel an electric current, the effect of electrons, and we can see images of physical objects transmitted to our consciousness as the effects of photons.

We are generally not aware of the flow of blood in our veins and arteries, or the flow of air in our lungs, and our bodies perform myriad functions of which we are generally unaware. Psychologists have identified mental functions associated with individual consciousness which we are blissfully unaware of in our normal waking state. Could it not be possible that we are also in touch with quantum reality on a level not normally available to our conscious minds? The answer undoubtedly lies in primary consciousness, which, as we proved in Chapter 2, had to exist prior to the appearance of the first physical particle. Primary consciousness, as the underlying principle, substance, and structure of the universe, nonlocal in time and space, is certainly in contact with our individualized consciousness, which is nonlocal over a limited region. In this indirect way, at least, we are in touch with all aspects of reality.

The Calculus of Distinctions, in so much as it reveals the structure of reality, may also reveal our relationship to the quantum world. If the thesis of this book is true: that the physical universe proceeds from consciousness, and not the other way around, then the explanation of both the microcosm and the macrocosm, mind and matter, inner and outer realities, incorporating the logical structure of quantum mechanics, relativity, and all other scientific theories, in so far as they are valid, may be found to already exist in the structure of primary consciousness.

Do quantum distinctions actually exist? Is the Copenhagen interpretation of quantum mechanics true? These questions have been asked by many scientists during the development of the current paradigm since Max Planck discovered the quantum nearly 100 years ago. These questions may never be answered to the satisfaction of everyone asking them, but it is important that we try. As we attempt to understand the nature of reality and the meaning of conscious existence, we realize that consciousness is the only thing we ever experience directly, and the question of the existence of discrete phenomena and the truth of finite ideas become less important than *the conscious realization of what we are*.

I believe that G. Spencer Brown realized this as he developed the Laws of Form. In his notes, at the end of the book, he says:

> There is a tendency, especially today, to regard existence as the source of reality, and thus as a central concept. But as soon as it is formally examined, existence is seen to be highly peripheral and, as such, especially corrupt (in the formal sense) and vulnerable. The concept of truth is more central, although still recognizably peripheral. If the weakness of present-day science is that it centres round existence, the weakness of present-day logic is that it centres round truth.
>
> Throughout the essay [the Laws of Form] we find no need of the concept of truth, apart from two avoidable appearances (true = open to proof) in the descriptive context. At no point, to say the least, is it a necessary inhabitant of the calculating forms. These forms are thus not only precursors of existence, they are also precursors of truth.
>
> It is, I am afraid, the intellectual block which most of us come up against at the points where, to experience the world clearly, we must abandon

> existence to truth, truth to indication, indication to form, and form to void, that has so held up the development of logic and its mathematics.
>
> ... A theorem is no more proved by logic and computation than a sonnet is written by grammar and rhetoric, or than a sonata is composed by harmony and counterpoint, or a picture painted by balance and perspective. Logic and computation, grammar and rhetoric, harmony and counterpoint, balance and perspective, can be seen in the work *after* it is created, but these forms are, in the final analysis, parasitic on, they have no existence apart from, the creativity of the work itself.

The existence of quanta as phenomena is less important than their existence as distinctions in consciousness, and the truth of any theory or paradigm is less important than the role it plays in opening a door to reality which may allow us to escape the limitations of our identification with individual body-mind consciousness and understand *who and what* we really are. Clearly, no distinction is relevant to a sentient being unless it can be registered in that being's consciousness, and now we are seeing that some form of consciousness is necessary for *any* distinction to exist. Thus the logical patterns and forms of distinctions are of primary importance to any new ontology.

This has been no more than a brief introduction to the Calculus of Distinctions. For a step-by-step development of the non-numerical basis of all mathematical systems, see Infinite Continuity, by this author, and Laws of Form by George Spencer Brown.

APPENDIX E, PART I

DERIVATION OF THE LORENTZ TRANSFORMATION EQUATIONS

It has been said that at one time, only a few, super-intelligent people were able to understand the theory of relativity. I don't know whether that was ever true or not. But if we restrict our concern to the Special Theory of Relativity, the subject of Einstein's 1905 paper, with the help of high school geometry, algebra, and a little imagination, the average person can understand it easily. Einstein's genius was in his ability to visualize physical situations involving motion, and to look beyond the accepted view of things.

Einstein knew that if you moved along a river bank parallel to the direction of motion of a wave traveling down the river, with the same speed as the wave, you would see the wave form as if it were standing still. He tried to imagine doing the same thing with an electromagnetic wave, a beam of light. What would it look like? Studying Maxwell's equation, he discovered that, no matter how fast you traveled, the light would always have the same speed relative to your point of view as an observer. This contradicted the everyday experience with moving objects. We can accelerate to the same speed as a train or plane, we can even imagine catching up with a speeding bullet, but not light. How could this be? Could there be something happening that only shows up when something is moving at or near the speed of light, an enormous speed relative to the speed with which most objects move in our range of experience? Einstein thought of the one thing that could explain it: time and space do not form a rigid framework within which everything moves. The measures of time and space change with relative motion.

Once this idea struck him, Einstein found that the mathematics needed to describe the phenomenon already existed. One important part of the mathematics had been developed by the Dutch physicist H. A. Lorentz in the process of explaining the apparent distortion of electron orbits due to their high rate of movement. The equations developed by Lorentz were just what Einstein needed to describe the distortion of time and space with motion. They are known as the Lorentz transformation equations. While Lorentz had derived them and used them specifically to correct what he perceived as observational error, Einstein saw them in the context of a universe in which time and space are relative and integrated them with Riemann's algebra and Minkowski's four-dimensional, complex-variable space to create a new paradigm. These transformation equations, so important to the basic understanding of relativity, may be derived by anyone from a simple geometric drawing (See Figure E-1.), using nothing more complicated than the Pythagorean theorem.

Imagine that a pulse of light is bouncing back and forth between two parallel mirrors that are on a spaceship moving with the velocity **v**, relative to the Earth. From the point of view of an observer at rest in the Earth reference frame, the light should take longer to go from **A** to **B** than from **A** to **C**, since the path **AB** is longer. But to an occupant of the spaceship, the paths **AC** and

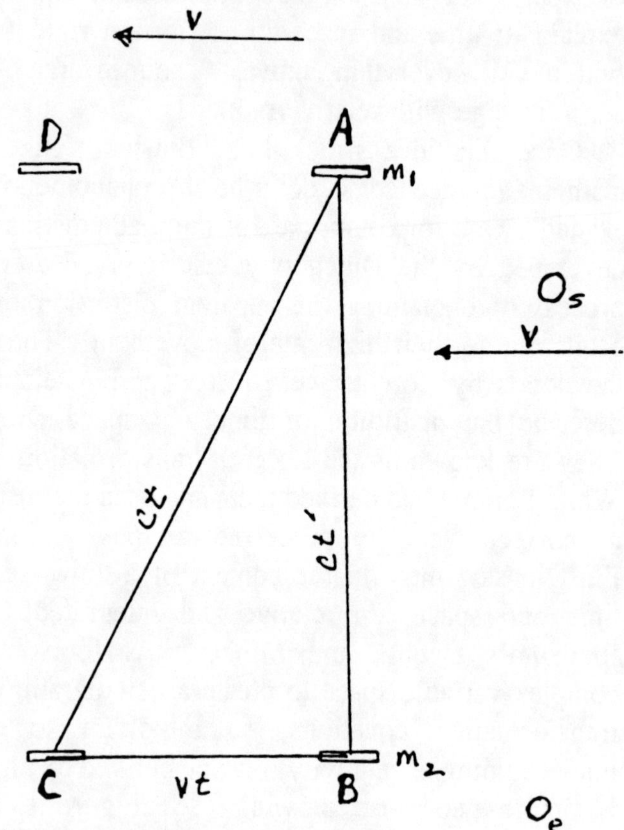

m_1, m_2 - mirrors on spaceship
A = Location of m_1 at Beginning of Observation
B = Location of m_2 at Beginning of Observation
C = Location of m_2 at time t After the Beginning of Observation
D = Location of m_1 at time t After the Beginning of Observation
O_e = Observer on Earth O_s = Observer on Spaceship
v = Velocity of O_s relative to O_e c = Speed of Light, t = Time Elapsed

Figure E-1

AB are the same. Thus, if the speed of light is the same for all observers, time on the spaceship, from the point of view of the observer on the Earth, will have to slow down by the factor $t/t' = AC/AB$. In order to find the value of this ratio in terms of the velocity, **v**, of the spaceship relative to the Earth, we may apply the Pythagorean theorem to the triangle formed by the points **ABC**, as seen from the stationary observer on the Earth. The pythagorean theorem states that "the sum of the squares of the legs of a right triangle is equal to the square of the hypotenuse." In other words, we may write:

$$(AC)^2 + (BC)^2 = (AB)^2 \qquad \text{Ex. E-1}$$

Dividing both sides of the equation by $(AB)^2$,

$$(AC/AB)^2 + (BC/AB)^2 = 1$$

or

$$(AC/AB)^2 = 1 - (BC/AB)^2,$$

and

$$AC/AB = \sqrt{1 - (BC/AB)^2} \qquad \text{Ex. E-2}$$

Looking at Figure E-1, we see that the distances are related to the velocities and times as follows:

$$AC/AB = t/t' \quad \text{and} \quad AC/AB = v/c$$

Substituting into Ex. E-2, we have:

$$t/t' = \sqrt{1 - (v/c)^2}$$

Inverting and transposing, we obtain:

$$t' = t/\sqrt{1 - (v/c)^2}, \qquad \text{Ex. E-3}$$

which is the familiar equation used to calculate relativistic time dilation.

The equation for contracting or shrinking yardsticks:

$$L' = L / \sqrt{1 - (v/c)^2} \qquad \text{Ex. E-4}$$

may be derived in exactly the same way, using the classical relationship between velocity, distance, and time and the Pythagorean theorem, using the appropriate ratios from the triangle in Figure E-1.

APPENDIX E, PART II

APPLICATION OF THE LORENTZ TRANSFORMATION EQUATIONS to OBSERVATIONS CONSIDERING PHOTONS AS MOVING OBJECTS

INTRODUCTION

The current scientific paradigm considers light to consist of complementary particle (photon) and wave phenomena, traveling in the near-vacuum of space at the constant speed **c**, relative to all observers. Observations made by moving observers can only be compared by using the Lorentz transformation equations derived in Part I of this Appendix.

Einstein rejected the idea of any causal link between consciousness and physical reality,[1] while Schrödinger, Bohr and Heisenberg declared that objective quantum phenomena do not exist until selected from probability functions by acts of observation. Bell's theorem,[2] predicts substantially higher correlations between physical properties of certain pairs of quanta for a "nonlocal" connected reality than Einstein's objective "local" reality. The Aspect experiment,[3] designed to determine the actual correlation of such quantum pairs, provides empirical evidence that a continuous, connected reality underlies quantum and high-speed phenomena. Relativity treats light as quanta of energy traveling as localized objects through every point of the space they traverse. Einstein's special theory of relativity requiring that these quanta always travel at a constant speed, relative to any observer's reference frame, regardless of the motion of the light source or the observer treats the quanta known as photons as if they were dimensionless points.

THE ROOT OF THE CONFLICT

The root of the conflict between relativity and quantum theory lies in their assumptions about the nature of light. Relativity describes light as wave energy with associated dimensionless quanta (photons), travelling through the vacuum of space at a constant speed relative to all observers. Quantum physics also recognizes the dual wave-particle nature of light, but insists that neither of these objective characteristics exists independent of measurement and observation. This difference is the root of the rift in the relativity-quantum model of reality. This rift, highlighted by the Einstein-Bohr debates of the 1930's, has never really been healed. It can only be healed by a theory that is consistent with both relativistic and quantum observations, including Bell's theorem and the Aspect experiment. In order to gain a clear understanding of the root conflict, we will have a detailed look at how the Lorentz transformations are applied under the differing assumptions of relativity and quantum theory.

For the moment, let's suppose that we've never heard of Einstein's shrinking yardsticks and slowing clocks, but we've been convinced by Maxwell's equations and empirical evidence that the speed of light in vacuum is constant for all observers, regardless of relative motion.

Now we'll visualize three reference frames: **K**, **O** and **K'** (Fig. 1), moving in such a manner that their origins coincide with a flash of light originating in **O** (Fig. 2). The space-time coordinates of the flash is (0,0,0) in all three reference frames. The **K** and **K'** reference frames are in continuous uniform movement relative to **O** at all times, before, at the time of, and after the flash. The relative velocity between **K** and **K'** is **v**, a large fraction of **c**, in a direction parallel to **x** and **x'** axes, so that these axes continue to be coincident and slide along each other as **K** and **K'** move apart after the flash (Fig. 3).

The motion of the light from the flash, along the **x** axis is described by the equation **x = ct** in the **K** reference frame, and by **x' = ct'** in the **K'** reference frame. These equations express the requirement of constant light speed relative to observers in both **K** and **K'** reference frames. The assumptions required for consistency are as follows:

Assumption 1: The speed of light is constant for all observers regardless of their relative motion.
Assumption 2: There is no preferred reference frame; i.e., all reference frames are physically equivalent.
Assumption 3: Light is an objective physical phenomenon, composed of waves and/or particles moving in space, independent of the consciousness of any observer.
Assumption 4: The coordinates of a point, p, in reference frame **K**, locating a wave front or photon originating from the flash in reference frame **O**, represents the same physical event as the point p' in reference frame **K'**.

Of these four assumptions, only the first two were stated explicitly by Einstein as the basis of relativity. However, the other two are implicit in the derivation of the Lorentz transformations. Einstein may simply have considered them too obvious to mention. He saw Assumption 3 as "the basis of all natural science." (See the quote[4] on page 14 in Chapter 1.) Assumption 4 is necessary unless we want to admit at the outset that the waves or particles of light moving in one reference frame are not the same waves or particles moving through the other. In fact, the tacit acceptance of Assumptions 3 and 4 is the basis of the conflict between Einstein's view of reality and that of Bohr and Heisenberg.

By inspection of Figure 3a, we see that x is greater than x' by the amount vt, the distance separating the origins of K and K' at time t, the time it takes the flash of light to reach $x = p$. Recognizing the fact that requiring $x = ct$ *and* $x' = ct'$ must affect the length of x' as observed from K, we write:

$$x = \alpha x' + vt \qquad (1)$$

where α represents the effect of the motion on x' as viewed from K. (Note that if it should happen that there is no effect, α will be equal to unity.) Similarly, from the point of view of K', (Figure 3c) paying attention to signs in the K' reference frame:

$$x' = \alpha x - vt \qquad (2)$$

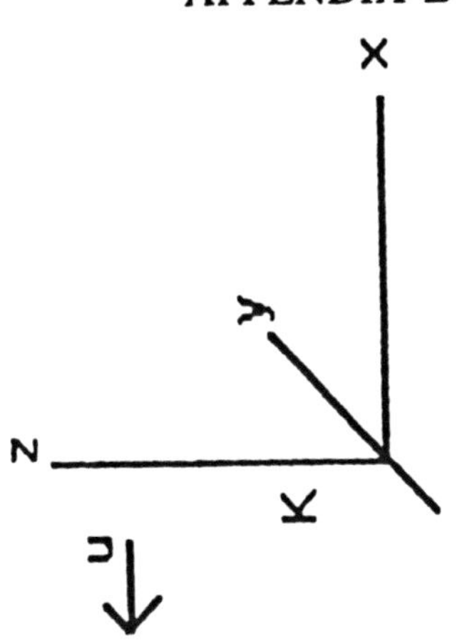

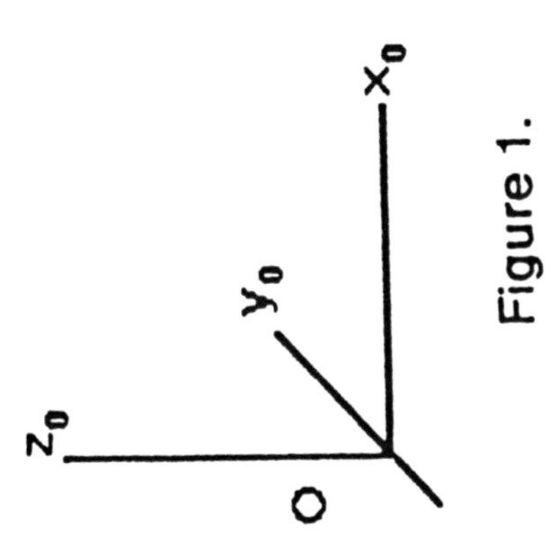

Figure 1.

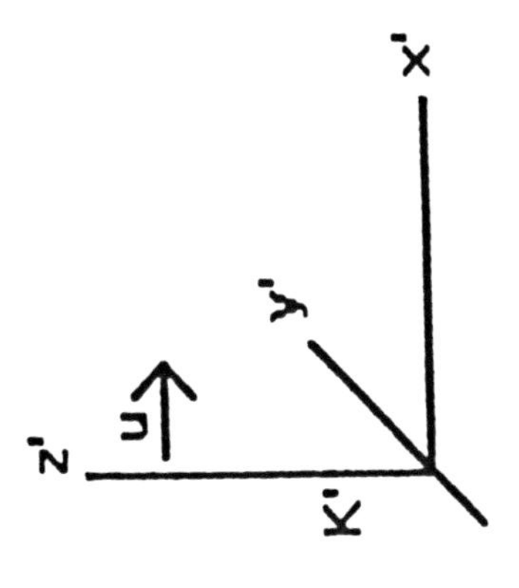

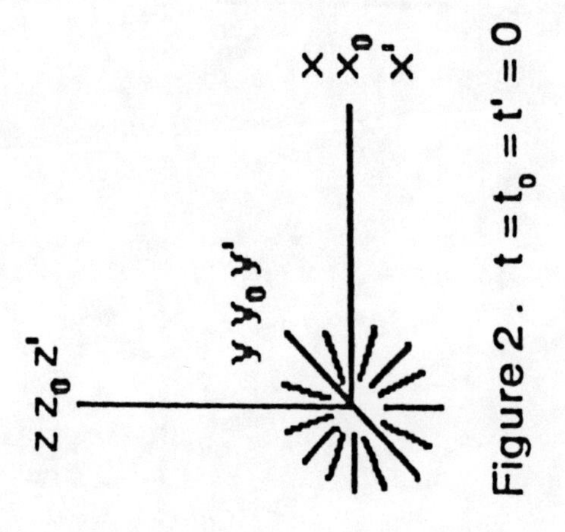

Figure 2. $t = t_0 = t' = 0$

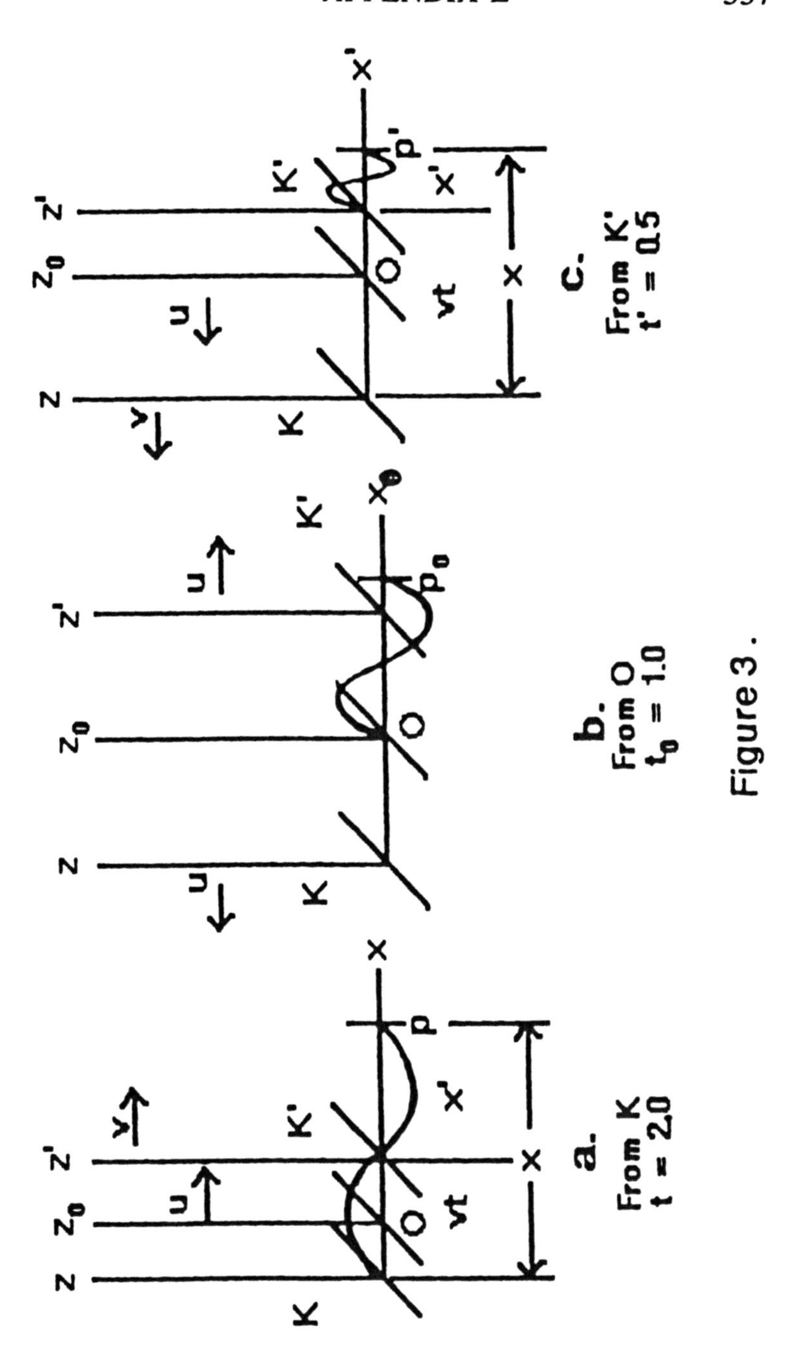
z
z0
z'
u
v
K
K'
O
vt
x
x'
p
a.
From K
t = 2.0
z
z0
z'
u
K
K'
O
x0
p0
b.
From O
t0 = 1.0
z
z0
z'
u
v
K
K'
O
vt
x
x'
p'
c.
From K'
t' = 0.5

Figure 3.

The factor α must be identical in (1) and (2), since by Assumption 2, the reference frames are equivalent. The sign in the right-hand side of the equation is negative since v is the relative velocity between K and K' and, if v is taken to be positive in K, it must be negative in K'.

Substituting x=ct and x'=ct'in (1) and (2) and solving for α, we have:

$$a=\frac{t}{t'}\left(\frac{c-v}{c}\right) \quad (3) \qquad \text{and} \qquad a=\frac{t'}{t}\left(\frac{c+v}{c}\right) \quad (4)$$

Equating the two expressions for α yields:

$$\frac{t}{t'}\left(\frac{c-v}{c}\right)=\frac{t'}{t}\left(\frac{c+v}{c}\right)\rightarrow\frac{t'}{t}=\sqrt{\frac{c-v}{c+v}} \quad (5)$$

Substituting (5) into (4), we get:

$$a=\sqrt{\frac{c-v}{c+v}}\left(\frac{c+v}{c}\right)=\sqrt{1-\frac{v^2}{c^2}} \quad (6)$$

Substituting this value of α into (1) and solving for x':

$$x'=\frac{x-vt}{\sqrt{1-\frac{v^2}{c^2}}} \quad (7)$$

Substituting (6) into (3) and solving for t' in terms of x and t, gives us :

$$t' = \frac{t - \frac{v}{c^2}x}{\sqrt{1 - \frac{v^2}{c^2}}} \quad (8)$$

Equations (7) and (8) are in the form of functions of x and t. Solving them for x and t, keeping in mind that v is positive in K and negative in K', we obtain:

$$x = \frac{x' - vt'}{\sqrt{1 - \frac{v^2}{c^2}}} \quad (9) \quad \text{and} \quad t = \frac{t' - \frac{v}{c^2}x'}{\sqrt{1 - \frac{v^2}{c^2}}} \quad (10)$$

Equations (7) through (10) are the Lorentz transformation equations for space-time coordinate systems in relative motion, as defined above. Under Assumptions 1 - 4, it is clear that the flash of light is considered to be an objective phenomenon moving along the x and x' axes, while time ranges from t=t'=0 until the wave and/or photon arrives at p in K and p' in K'. Quantum physics, however, per the Copenhagen interpretation, does not agree with this, but instead, holds that no separate quantum phenomena exist until a measurement or observation is made. Thus there is a clear conflict. Derivation of the

Lorentz transformation equations depends upon the continuous existence of a separate objective phenomenon from (x,t)=(x',t')=(0,0) to p, while quantum theory cannot allow this. During the Einstein-Bohr debates, Einstein, Podolsky and Rosen published a paper entitled "*Can the Quantum Mechanical Description of Physical Reality be Considered Complete?*"[5] Starting with the basic assumptions of relativity, they produced what became known as the EPR paradox, refuting Heisenberg's uncertainty principle. Bohr rejected their argument on the basis that they had made an unwarranted assumption: the assumption that elementary particles exist before they are measured or observed. Einstein never accepted this idea because it implied a causal link between consciousness and objective physical reality and a nonlocal connection between distant objects, violating assumptions 1 and 3 above. Bell's theorem and the empirical evidence of the Aspect experiment, however, have upheld Bohr's argument.

AN EXPERIMENT

In order to see how this basic conflict arises in application, and how Transcendental Physics resolves it, we will investigate measurements and observations made in the moving inertial frames of Figures 1 - 3. We begin our experiment at a time prior to the flash of light, when we find the observers of K and K' at rest relative to the O reference frame. At that time, they synchronize their clocks and agree upon the following procedures, which, for the sake of this discussion, we will say they have the technology to accomplish:

a) Their measuring instruments will all be calibrated in the following basic units: The basic unit of length will be a specific wave length of light contained in the flash of light as viewed from the rest frame, **O**. The basic unit of time will be the time required for one unit-length wave to pass a stationary point in the **O** reference frame. This may be called the *transit time* for the standard unit wave length. (Note that in this system of units, $\mathbf{c}=1$.)

b) The K and K' observers will then accelerate from the **O** rest frame in opposite directions turning at some point to approach the point (0,0,0) in the **O** reference frame at equal and opposite speeds. They will each measure wave lengths and transit times, starting at the instant of the flash in O, and use the Lorentz transformations to determine the coordinates of these measurements of the light wave in the other observer's reference frames.

c) K and K' will each be travelling at the speed **u**, relative to the original inertial frame **O**, approaching each other, according to the theory of relativity at the combined speed of $\mathbf{v}=2\mathbf{u}/(1+\mathbf{u}^2)$. They will maintain constant velocities so that after the origins of their reference systems coincide, they will be separating at the relative speed **v**.

d) They will record their observations for later comparison.

As K and K' approach $\mathbf{x_0}=\mathbf{t_0}=0$, where all three coordinate systems will coincide at the time of the flash, the observers will be readying their equipment to measure or calculate the following:

1) The other's basic units of length and time relative to his own,
2) The standard wave length in his and the other's coordinate systems, and
3) The transit time of the observed light wave in each of the coordinate systems.

After the flash, observers in each of the reference frames will see the other two receding. For example, an observer in **K** will see **O** dropping away at the speed **u**, and K' moving away on the other side of **O** at the speed **v**. From **K**, the basic units of length in the other two reference frames will be shorter than his, and their clocks will run more slowly.

Since the flash of light originates at the coincident origins of the three reference frames, the coordinates of one end of the light wave will be the same for all three: $(x,t)=(x_0t_0)=(x',t')=(0,0)$. The other end will be at $p_0=(1,1)$ in the **O** reference frame, and we can use the Lorentz transformations to determine the corresponding coordinates in **k** and **K'**, as seen from **O**. The length of the wave in **k** and **K'**will be equal to the differences between the locations of the ends of the wave. So, from **O**'s point of view:

$$\lambda_K=x(1,1)-x(0,0) \qquad \text{and} \qquad \lambda_{K'}=x'(1,1)-x'(0,0)$$

where $x(1,1)$ represents the value of equation (9) when x' is replaced by x_0, t' by t' by t_0 and $(x_0,t_0)=(1,1)$. And $x'(1,1)$ represents the value of equation (7) with x replaced by x_0 and **t** by t_0. So we have:

$$\lambda_K = \frac{1+u\times 1}{\sqrt{1-u^2}} - \frac{0+u\times 0}{\sqrt{1-u^2}} = \sqrt{\frac{1+u}{1-u}} \qquad (11)$$

$$\lambda_{K'} = \frac{1-u\times 1}{\sqrt{1-u^2}} - \frac{0-u\times 0}{\sqrt{1-u^2}} = \sqrt{\frac{1-u}{1+u}} \qquad (12)$$

The final expressions of equations (11) and (12) are the well-known expressions for Doppler effects in light waves. Data collected by each observer are presented in Tables 1-3. The values given are for u= 0.6c.

Table 1: Data From K

Physical Parameters	Reference Frame		
	K	O	K'
Unit Length	1.00	0.80	0.47
Unit Time	1.00	0.80	0.47
Transit Time	2.00	1.60	0.94
Wave Length	2.00	1.60*	0.94*

Table 2: Data From O

Physical Parameters	Reference Frame		
	K	O	K'
Unit Length	0.80	1.00	0.80
Unit Time	0.80	1.00	0.80
Transit Time	0.80	1.00	0.80
Wave Length	0.80*	1.00	0.80*

Table 3: Data From K'

Physical Parameters	Reference Frame		
	K	O	K'
Unit Length	0.47	0.80	1.00
Unit Time	0.47	0.80	1.00
Transit Time	0.24	0.40	0.50
Wave Length	0.24*	0.40*	0.50

*Since each observer sees only one light wave, the wave lengths in the other inertial frames are determined using clock times calculated by Lorentz time transformations. All other values in the tables, under the assumptions of our experiment, are observed or measured

ANALYSIS OF RESULTS

The information in Tables 1-3 represents data that would be obtained from instantaneous snapshots,i.e., neglecting the travel time for light to reach the cameras. Each observer sets his camera to take a snapshot at $t=t_0=t'=0$, again when his clock indicates that one time unit has passed and at the end of one wave cycle, if that is different than one time unit on his clock.

By definition, the wave length equals one unit in O. From equations (11) and (12), we know that the wave lengths observed by K and K' will be:

$$\lambda_K = \sqrt{\frac{1+0.6}{1-0.6}} = 2.00$$

$$\text{and } \lambda_{K'} = \sqrt{\frac{1-0.6}{1+0.6}} = 0.50$$

In our system of units, defined so that $c=1$, $\lambda=t$. with $u=0.6c$, all cameras will snap at $t=0$; camera K will snap again at $t=1.00$ and at $t=2.00$; camera Oat $t_0=1.00$; and camera K' at $t'=0.50$ and 1.00.

The snapshot taken at $t=0$ by the camera in K will show the unit of length in K' lying along the x axis from $x'=0$ to $x'=1$. The length of the unit in K' photographed from K is given by:

$$L_{K'}=x(1,0)-x(0,0) \tag{15}$$

Using equation (7) and recalling that $c=1$ in our system of units, we have:

$$x=x'\sqrt{1-v^2}+vt \tag{16}$$

Since $t=0$, equation (15) yields:

$$L_{K'}=\sqrt{1-v^2}-0=0.47 \tag{17}$$

Similarly, K will find the unit length in O to be:

$$L_O=\sqrt{1-u^2}=0.80 \tag{18}$$

When the clock in K reads 1, the K camera will photograph the clock in K' reading t'=0.47. This conforms to equation (8);

$$t=t'\sqrt{1-v^2}+vx \tag{19}$$

Substituting (19) into $\Delta t'_K=t(1,0)-t(0,0)$ (20) we get:

$$\Delta t'_K=\sqrt{1-v^2}+vx-(0+vx)=\sqrt{1-v^2}=0.47 \tag{21}$$

Similarly, the "snapshot" from K will show the time unit in O to be:

$$\Delta t_O=\sqrt{1-u^2}=0.80 \tag{22}$$

Equations (15) through (22) are consistent with Einstein's discussion of the behavior of measuring rods and clocks with motion.[6]

We have determined that for every time unit that passes in his reference frame, K will see 0.80 and o.47 units passing in O and K', respectively. Consequently,since **u** and **v** are uniform velocities, the photo taken from K when t=2 will show the wave reaching the point p and the clocks in the K' and O reference frames reading 0.94 and 1.60, respectively. The observer in K will record these readings and conclude that the corresponding wave lengths in the other reference frames are $\lambda_O=1.60$ and $\lambda_{K'}=0.94$, since he knows that the wave has reached the point p and c=1 implies that wave length equals transit time in any reference frame.

The observers in O and K' proceed similarly. The results are recorded in Tables 1 through 3. We find, however, that there is a discrepancy between the wave lengths corresponding to the photographed clock times recorded by the observers and the wave lengths consistent with the measurement made by each observer in his own reference frame and transformed with the Lorentz equations. The wave lengths for O and K', consistent with an observed wave length of 2.00 in K are:

$$\lambda_O = \sqrt{\frac{1-u}{1+u}}\lambda_K = 0.5\times 2.00 = 1.00$$

and $$\lambda_{K'} = \sqrt{\frac{1-v}{1+v}}\lambda_K = 0.25\times 2.00 = 0.50$$

But if we replace the values of the parameters marked with asterisks in the tables with these values, it becomes clear that the photographed clock times cannot be transit times, since c=1 implies that wave length = transit time.

Given the coordinates of a point in one reference frame, the Lorentz transformation equations are supposed to yield the coordinates of that point in a second reference frame that is in uniform motion relative to the first frame of reference. But we see that the three points p, p_o and p' are not representative of the same physical event in space-time, and ***the three observers in our experiment cannot be observing and measuring the same light wave***. The data indicate that, because of their motion relative to one another, the observers are seeing and measuring three different light waves. If we follow the same line of reasoning while replacing the wave length with a photon assumed to be traversing the wave-length distance in one unit time, we come to the same conclusion: Three observers in relative motion detect three *different* photons.

If we replace the light wave/photon with an ordinary object, like a baseball, traveling from the origin to point p in the **O** reference frame, and go through the same calculations, replacing the parameter "wave length" in each table by the distance traveled by the object, the problem goes away. Transformation of the observed distance traveled in one reference frame to the other reference frames does not conflict with any expectation, since we know that the object is traveling at different speeds in the other reference frames.

These results clearly contradict Assumption #4 which says that a wave or photon of light is a single physical event that may be observed from any reference frame regardless of relative motion. The Copenhagen interpretation of quantum mechanics resolves this problem nicely: Elementary phenomena such as the photon simply do not exist until a quantum of energy registers on a receptor. Thus the electromagnetic energy moving from the origin to point p can only be described by a probability wave function until the wave is "collapsed" to one of its possible states by a measurement or observation made by an observer.

REFERENCES

1. F. David Peat, Einstein's Moon, Contemporary Books, Inc., Chicago IL, 1990, pp. 68-69.
2. Op. cit., pp. 85-117.
3. Ibid., pp. 117-122
4. Albert Einstein, in: James Clerke Maxwell: A Commemorative Volume, Cambridge University Press, 1931.
5. Albert Einstein, Boris Podolsky and Nathan Rosen, *Can the Quantum Mechanical Description of Physical Reality Be Considered Complete?*, Physical Review, 1935.
6. Albert Einstein, Relativity, The Special and General Theory, Crown Publishers Inc., New York, 1961, pp.35-37.

SELECTED BIBLIOGRAPHY

Barrow, John and Joseph Silk, The Left Hand of Creation, Basic Books, New York, 1983. Non-technical treatment of the origin and evolution of the expanding universe

Brown, G. Spencer, Laws of Form, George Allen and Unwin Ltd., London, 1969, The Julian Press, New York, 1977

Bohm, David, Wholeness and the Implicate Order, Routledge and Kegan Paul, London, 1981. Theory explaining nonlocal phenomena

Capra, Fritjof, The Tao of Physics, Shambhala, Boulder, Colorado, 1976, Bantam Books, New York, 1977. Parallels and similarities between physics and mysticism

Casti, John L., Paradigms Lost, William Morrow and Company, New York, 1989. Six major mysteries confronting modern science

Close, Edward R., The Book of Atma, Libra Publishers, New York, 1977. A discussion of meditation, reincarnation and Cosmic Consciousness

Cramer, John G., "Generalized Absorber Theory and the Einstein-Podolsky-Rosen Paradox", Physical Review, Vol. 22, 1980

_______, "The Transactional Interpretation of Quantum Mechanics", Reviews of Modern Physics, 1986

Davies, Paul, God and the New Physics, Simon and Schuster, New York, 1984. A look at the impact of science upon what were formerly considered as religious issues

_______, The Cosmic Blueprint, Touchstone, Simon and Schuster, New York, 1988.
Arguing for a self-organizing reality

Einstein, Albert, Ideas and Opinions, Bonanza Books, Crown Publishers, New York, 1954, Einstein's ideas and opinions on science and religion

_______, Relativity, the Special and the General Theory, Crown Publishers, New York, Fifteenth Edition, 1961, ⌐ by the Estate of Albert Einstein

Ferris, Timothy, The Red Limit, second edition, William Morrow and Company, New York, 1983. Historical account of astronomy and cosmology focusing on the discovery of evidence of the expanding universe

Friedman, Norman, Bridging Science and Spirit, Living Lakes Books, St. Louis, MO, 1994. Common elements in David Bohm's physics , the perenial philosophy and Seth

Goswami, Amit, The Self-Aware Universe, G.P. Putnam's Sons, New York, 1995. The case for scientific idealism

Gribben, John and Martin Rees, Cosmic Coincidences, Bantam Books, New York, September, 1989. Evidence of design in the universe and the meaning of seeming coincidences that allow us to exist

Hawking, Stephen H., A Brief History of Time, Bantam Books, New York, 1988. A review of cosmological theories, asking questions like: was there a beginning of the Universe? Will there be an end?

Hofstadter, Douglas R., Metamagical Themas: Questing for the Essence of Mind and Pattern, Basic Books, New York, 1985. Relating literary, scientific and artistic studies to our search for meaning.

_______, Gödel, Escher, Bach: An Eternal Golden Braid, Basic Books, New York, 1981. Exploring the meaning of self-referential systems or "strangeloops" of art, science and music

Pagels, Heinz R., Perfect Symmetry, Bantam Books, New York, 1985. An extremely well-written guide to the frontier of cosmology and physics

Peat, F. David, Einstein's Moon, Contemporary Books, Chicago, 1990. A clear discussion of Bell's theorem and the quest for quantum reality

Rucker, Rudy, Infinity and the Mind, Bantam Books, New York, 1983. A consideration of the meaning of man and mind on the backdrop of an infinite reality

Sheldrake, Rupert, The Presence of the Past, Random House Vintage Books, New York, 1989. The author presents the thesis that nature evolves because it possesses memory, which he calls morphic resonance, repetition in time

Strauch, Ralph, The Reality Illusion, the Theosophical Publishing House, Wheaton, IL, 1983. Based on the thesis that we create the world we experience

Talbot, Michael, Beyond The Quantum, Bantam Books, New York, 1988. A discussion of God, Psychic Phenomena, Reality and consciousness in scientific revolution

Trefil, John, The Moment of Creation, Charles Schreibner & Sons, New York, 1983. An attempt to develop a comprehensive worldview from the basic ideas behind modern physics and the big bang theory

Wallace, B. Alan, Choosing Reality, A Contemplative View of Physics and the Mind, New Science Library, Shambhala Publications, Boston, Massachusetts, 1989. An exploration of the relationship of the reality we experience to the reality revealed by science.

Wheeler, John A., A.R. Marlow, editor, The Mathematical Foundations of Quantum Mechanics, Academic Press, New York,, 1978.

Wheeler, John A., At Home in the Universe, American Institute of Physics, 1994. Quantum mechanics, time, space, and our participatory universe

_______, Delayed-Choice Experiments and the Bohr-Einstein Dialogue, The American Philosophical Society, New York, 1980.

Wilber, Ken, Quantum Questions, New Science Library, Shambhala Publications, Boulder, Colorado, 1984. A collection of mystical writings of some of the world's great physicists with commentary by Ken Wilber, concluding that science and mysticism are both valid approaches to separate domains of reality

Wolf, Fred Alan, Parallel Universes, Simon and Schuster, New York, 1988. The case for the existence of parallel universes

_______, Taking the Quantum Leap, Harper and Row, New York, Perennial Library Edition, 1989. An updated edition of his 1981 book wherein he asserts that quantum theory is still valid and "as weird as ever".

GLOSSARY

Assumptions: Underlying concepts thought to be self-evident or unprovable. An underlying assumption of scientific materialism, for example, is that physical reality exists independent of consciousness.

Calculation: The logical transformation of one form or expression into another equivalent but different form.

Calculus: A logical system developed for the purpose of the logical transformation of forms (calculation).

Calculus of Distinctions: A mathematical system developed for the purpose of describing reality in terms of logical distinctions.

Complementarity, Principle of: Wave and particle phenomena complement each other, since both are necessary to explain experimental observations. Other examples include structure and entropy, consciousness and matter, relativity and quantum theory.

Content of a Distinction: The value assigned to a distinction.

Distinction: The differentiation of any region or part from the rest of the universe.

Doppler Effect: The lengthening of wavelength due to relative motion.

Entropy: (Statistical mechanics) Disorder. The even-ness of the distribution of energy in a closed system.

Existential Distinction: A distinct entity or object that exists independent of a given conscious observer. The geometry of a conceptual reference frame, undistorted by any force field, may be termed existential.

Expression: A symbolic representation of a distinction or group of distinctions in the form of a logical statement expressing an equivalence or other relationship.

Extent: Distinctions of space and time.

Extra-Spatial Dimensions: Measure of extent other than the three dimensions of apparent space.

Extropic: Contracting and organizing force.

Extropic-Entropic: Opposing forces of order and disorder, contraction and expansion.

Extropy: The contracting and organizing force reflected by structure and order; the opposite of entropy.

Field: The distribution of the substance of reality, in the forms of matter, energy and consciousness, in the multi-dimensional space-time continuum.

Forms of Substance: The three forms of universal substance are consciousness, energy and matter.

Gestaltenraum: Non-quantum space in consciousness where images corresponding to objective reality are formed.

Hilbert Space: Multi-dimensional space. A mathematical system for handling functions of n variables (n= any number from one to infinity) as mutually othogonal dimensions (eg. three dimensions of space plus one of time). Developed by the German mathematician David Hilbert.

Individualized Consciousness: Limited nonlocal consciousness associated with a finite form.

Infinite Continuity: The property of consciousness which allows for nonlocality and endless division.

Initial Expressions: The initial expressions of the Calculus of distinctions; pages 320 & 321.

Light: Electromagnetic radiation, existing in undifferentiated probabilistic wave form until a measurement or observation is made.

Local Reality: A reality with no superluminal interaction between objects separated in space.

Lower Limit Horizon (LLH): The micro-scale horizon beyond which no smaller direct measurement is possible.

Memory Function: An oscillatory calculus of distinctions expression which remembers, in the current form, the value of previous states.

Morphic or Morphogenetic field: An energy field, created by memories or reflexive patterns of past and future states, which guides the evolution of forms.

Nexus: Interface between consciousness and matter.

Nonlocal Reality: A reality in which apparently separate regions are connected by instantaneous interactions.

Non-Numerical Mathematics: The mathematics dealing with calculations prior to separation and enumeration.

Non-Quantum Receptor: The necessary final receptor implied by the Copenhagen interpretation of quantum mechanics to bring physical reality out of the probability wave function of possible states

Original Consciousness: The agency of extropy and the non-quantum receptor of the primary distinction.

Partition: Two or more distinctions of the same order within a region.

Participatory Reality: A reality in which consciousness participates as a formative agent.

Perceptual Distinction: A distinction, the form of which depends upon conscious observation. Observations related to an arbitrary reference frame and limited by a physical apparatus to perception of a finite range of radiant energy are perceptual distinctions.

Primary Distinction: The first distinction drawn in the Void.

Primary Expressions: The primary symbolic representations of the logical relationships between distinctions.

Region: The extent, in three or more dimensions, over which a distinction operates.

Self-Referential Reality: All-inclusive reality, outside of which no reference is possible.

Simplification: The value of an expression is equivalent to the value to which it can be simplified.

Sub-region: Distinction within a region already distinguished.

Timeline: A one-dimensional trace selected out of the three-dimensional time continuum by consciousness in order to limit and focus observation.

Transcendental Physics: The applications of the calculus of distinctions to the description of the reality brought out of the continuous void by the drawing of distinctions; transcends classical and modern physics by including consciousness and its interaction with the mass/energy - space/time continuum.

Transcendental Value or Statement: A value or statement that is neither true, false, nor meaningless in the context of standard logic and three-dimensional space, but is significant and necessary in the description of reality.

Universal Substance: That which manifests as mind, matter and/or energy; the basis of existential reality.

Upper Limit Horizon: An extremely large region of physical space defined by the sphere of all that may be perceived by individualized consciousness. This horizon is determined by the limitation of physical perception.

Value of a Distinction: The state of the region distinguished, i.e., either distinct or void.

Vanishing Point: Points defined by the LLH and ULH where finite increments of a variable or variables vanish from perception.

Void, The: The infinite expanse of undifferentiated substance, prior to the primary distinction.

INDEX

Printed in the United States
125936LV00003B/15/A